SIBLING SPECIES OF TITMICE
IN THE *PARUS INORNATUS* COMPLEX
(AVES: PARIDAE)

Sibling Species of Titmice in the *Parus inornatus* Complex (Aves: Paridae)

Carla Cicero

A Contribution from the Museum of Vertebrate Zoology of the University of California at Berkeley

UNIVERSITY OF CALIFORNIA PRESS
Berkeley • Los Angeles • London

UNIVERSITY OF CALIFORNIA PUBLICATIONS IN ZOOLOGY

Volume 128
Issue Date: January 1996

UNIVERSITY OF CALIFORNIA PRESS
BERKELEY AND LOS ANGELES, CALIFORNIA

UNIVERSITY OF CALIFORNIA PRESS, LTD.
LONDON, ENGLAND

Library of Congress Cataloging-in-Publication Data

Cicero, Carla, 1959–
    Sibling species of titmice in the Parus inornatus complex
(Aves:Paridae) / Carla Cicero.
        p.   cm. — (A Contribution from the Museum of Vertebrate
Zoology of the University of California at Berkeley) (University of
California publications in zoology ; v. 128)
    Includes bibliographical references.
    ISBN 0-520-09808-0 (pbk. : alk. paper)
    1. Plain titmouse.   I. Title. II. Series. III. Series:
University of California publications in zoology ; v. 128.
QL696.P2615C535   1996
598.8'24—dc20   95-46527
                                                           CIP

The paper used in this publication meets the minimum requirements
of American National Standard for Information Sciences—Permanence of
Paper for Printed Library Materials, ANSI Z39.48-1984.

In memory of my mother,

whose support and encouragement made this work possible

Contents

Figures

Tables

Acknowledgments

This work would not have been possible without the assistance of numerous people from the University of California and other institutions. I am particularly indebted to Ned K. Johnson, whose advice, teaching, and continual encouragement were instrumental to this project. Alan J. Baker, Keith L. Dixon, Thomas O. Duncan, Ned K. Johnson, William J. Libby, and Joe R. McBride all provided critical comments that substantially improved the manuscript. Harry W. Greene, James L. Patton, and David B. Wake also contributed through their courses, papers, and interactions, which enhanced my understanding of geographic variation and evolutionary biology.

Monica F. Frelow, Enrique P. Lessa, Jill A. Marten, Christian Orrego, and Margaret F. Smith provided expert assistance in the Evolutionary Genetics Laboratory of the Museum of Vertebrate Zoology (MVZ). Robert E. Jones supplied ammunition and other materials necessary for the collection of specimens. William Ramsey designed and manufactured auxilliary barrels for specimen collecting. Jeffrey G. Groth collected three specimens from New Mexico, and Ned K. Johnson assisted with tape-recording and collecting of specimens incidentally to other field work. Christopher A. Meecham and Alan G. Stangenberger helped with computer analyses. Jim Ashby of the Desert Research Institute provided climatological data for the western United States. Barbara R. Stein assisted with curation and database management of specimens. Karen Klitz helped in numerous ways with the figures. Rose Anne White of the University of California Press advised on preparation of the final camera-ready copy of this monograph.

I appreciate the assistance of all curators and curatorial staff of museums from which specimens were borrowed (see "Materials and Method"). Steven F. Bailey and Luis F. Baptista of the California Academy of Sciences, and George F. Barrowclough of the American Museum of Natural History, provided access to the avian collections in their care. Steven W. Cardiff, Donna L. Dittman, and Robert M. Zink (Louisiana State University Museum of Natural Science) sent critical tissue of *Parus inornatus* from the eastern Mojave Desert.

Scientific collecting permits were issued by the U.S. Fish and Wildlife Service, U.S. National Park Service (Joshua Tree National Monument), California Department of Fish and Game, Arizona Game and Fish Department, Nevada Department of Wildlife, Utah Division of Wildlife Resources, Colorado Division of Wildlife, Oregon Department of Fish and Wildlife, and New Mexico Department of Game and Fish. Additional authorizations for collecting were provided by: Mark Stromberg, Hastings Natural History Reservation, Monterey County, California; Michael Connor, Sierra Foothills Range Experiment Station, Yuba County, California; Ray Krauss, Homestake Mining Company McLaughlin Mine site, Yolo County, California; John Arnold, private ranch, San Luis Obispo County, California. John P. Hubbard of the New Mexico Department of Fish and Game provided information on places to collect in New Mexico. The Animal Care and Use Committee on the Berkeley campus of the University of California approved the necessary protocols.

This study was funded primarily by a National Science Foundation Dissertation Improvement Grant (BSR-9001120) to Ned K. Johnson, Principal Investigator. The American Ornithologists' Union, Museum of Vertebrate Zoology, and Sigma Xi (National Chapter) provided additional financial grants and awards. Joe R. McBride assisted with the purchase of digital calipers and a computer interface for the collection of morphometric data.

INTRODUCTION

Because of the importance of geographic variation to microevolutionary theory (Gould and Johnston 1972; Zink and Remsen 1986; Zink 1989), intraspecific studies have served as the basic framework for developing and testing major ideas in the speciation of natural populations (Mayr 1942, 1963; Selander 1971; Bush 1975; Endler 1977; Templeton 1980, 1981; Barton and Charlesworth 1984; Carson and Templeton 1984; Otte and Endler 1989; Coyne 1992). This research has greatly advanced our understanding of such important evolutionary processes as hybridization (Barton and Hewitt 1985; Littlejohn and Watson 1985; Hewitt 1988; Arnold 1992), gene flow (Barrowclough 1980a; Slatkin 1985a, 1987; Rockwell and Barrowclough 1987), adaptation (Johnston 1969, 1972; J.F. McDonald 1983; Mousseau and Ruff 1989), environmental induction (Straney and Patton 1980; James 1983; Patton and Brylski 1987), Pleistocene disjunction (Mengel 1964; Hubbard 1973; Mayr and Ohara 1986; Riddle and Honeycutt 1990), and the evolution of reproductive isolating mechanisms (N.K. Johnson 1980; Walsh 1982; Frost and Platz 1983; M.C. Baker and A.E.M. Baker 1990). While avian taxa have provided an especially rich source of material for evolutionary studies, thorough surveys of geographic variation in birds are relatively scarce. A major problem with many studies is the spotty geographic coverage (Zink and Remsen 1986). This is especially true for intraspecific studies of genetic variation which, although growing in number, are still fairly uncommon (for reviews, see Corbin 1983; Barrowclough et al. 1985; Barrowclough and Johnson 1988). Another limitation of intraspecific studies is that most workers have examined variation in only one or two suites of traits. Few systematists (e.g., N.K. Johnson 1980; Atwood 1988; Pitochelli 1990; Bell 1992; Groth 1993) have attempted to integrate data from more than two sets of characters (morphology, coloration, genetics [allozymes, DNA], ecology, behavior, and/or voice) over the breeding range of the species being studied.

1

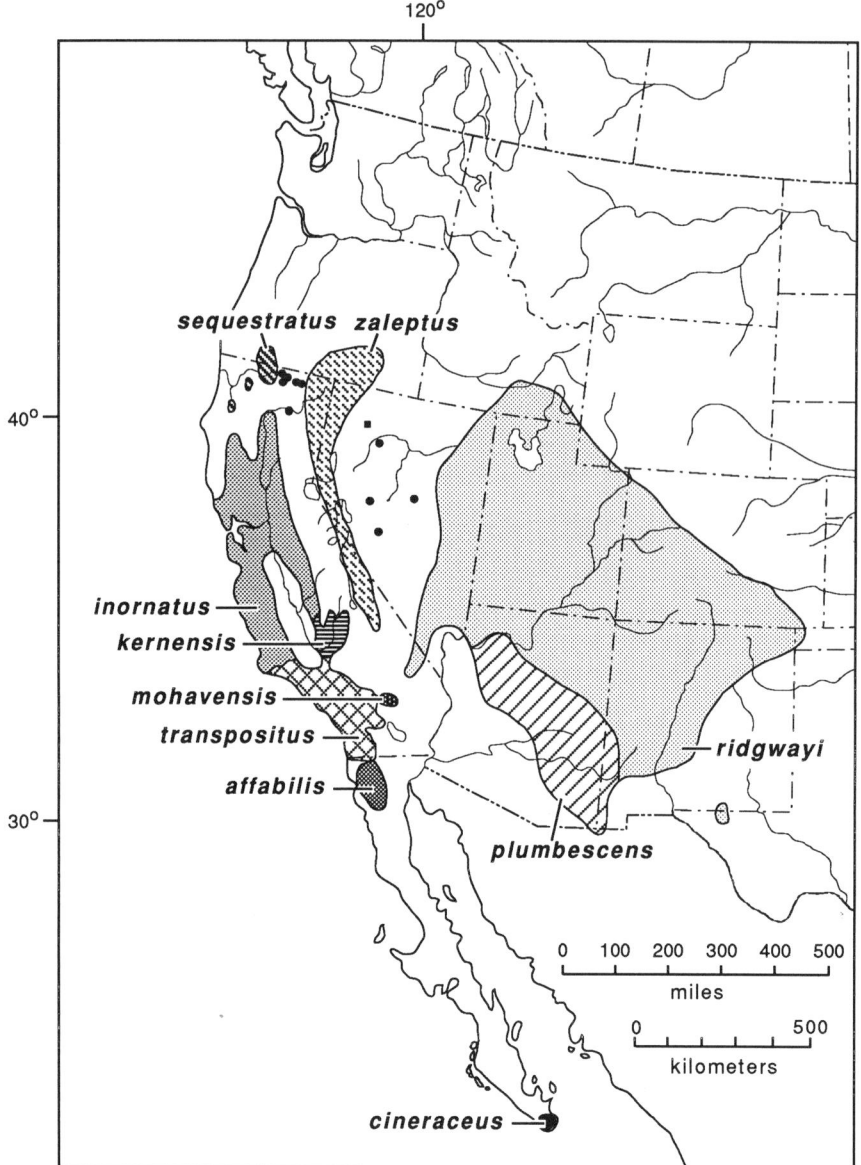

FIGURE 1. Generalized distribution of the 10 currently recognized subspecies of *Parus inornatus* in western North America. Outlined ranges are based primarily on Grinnell and Miller (1944:308) and the American Ornithologists' Union (1957). Symbols indicate new records obtained during this study (dots = specimen records; square = sight observation). Note the isolated distributions of *sequestratus, zaleptus,* and *cineraceus.* Populations, especially in the interior (*zaleptus, ridgwayi, plumbescens*), are not necessarily continuous within the outlined areas.

THE *PARUS INORNATUS* COMPLEX

Most comprehensive studies of intraspecific variability in birds have focused on taxa that are highly variable phenotypically (e.g., *Eremophila alpestris* [Behle 1942]; *Zonotrichia leucophrys* [Banks 1964]; *Passerella iliaca* [Zink 1986]), that provide cases of rapid adaptive evolution (e.g., *Passer domesticus* [Johnston and Selander 1964, 1971]), and/or that exemplify ecomorphologic variation on a broad geographic scale (e.g., *Agelaius phoeniceus* [Power 1970]; *Melospiza melodia* [Aldrich 1984]; *Turdus migratorius* [Aldrich and James 1991]). Phenotypically similar sibling species (e.g., *Empidonax* [N.K. Johnson 1963, 1980]; *Sitta* [Norris 1958]; *Polioptila* [Atwood 1988]) have also been the subject of numerous investigations directed at questions pertaining to species limits and reproductive isolation. The lack of *striking* phenotypic differentiation in the Plain Titmouse (*Parus inornatus*) complex is probably a major reason why this species has failed to attract the attention of avian systematists. Nonetheless, the species has several intriguing attributes that deserve modern systematic treatment.

Restricted to western North America, the range of *P. inornatus* is subdivided into several geographically isolated groups of populations that extend from southern Oregon to Baja California and from the western Great Basin to the Rocky Mountains and southwestern United States (Fig. 1). These groups are separated by areas of unsuitable habitat which represent formidable barriers for this *strictly resident* species. In the west, cool coniferous forest isolates populations on either side of the Sierra Nevada and the Cascade Mountains. Similar habitat near Mt. Shasta separates populations in northern California from those at the head of the Sacramento Valley to the south. Populations in the Great Basin and Southwest occupy disjunct distributions on the lower slopes of mountain ranges that are isolated by cold or warm desert environments, respectively. Desert habitat also separates populations in northern Baja California from the strongly isolated population in the Cape District of southern Baja California.

Despite its name, previous workers saw sufficient variability to describe 11 subspecies of *P. inornatus*, 10 of which are currently recognized by the American Ornithologists' Union (1957; Fig. 1). These forms are identified on the basis of relatively subtle differences in overall size, bill size and shape, and plumage color (Table 1 and Fig. 2). Perhaps of greater interest are the strong differences in ecology, distribution, and density that characterize allopatric and phenotypically divergent groups of populations. Habitat associations vary generally from oak-dominated woodland along the Pacific slope to juniper or pinyon-juniper woodland in the Great Basin and Southwest. Because of the disjunct distribution of woodland in the interior, titmice there occur as fragmented populations with relatively little contact between them. Habitats and populations are more continuous on the Pacific slope, and densities also are notably higher in that region. These geographic differences in population density and distribution in relation to habitat preference have important implications for interpreting patterns of genetic variation and gene flow, especially given the sedentary behavior of the species.

TABLE 1. Current classification (American Ornithologists' Union 1957) and comparative descriptions of subspecies in the *Parus inornatus* complex.

Current Subspecies	Subspecies Comparison[a]	Color	Overall Size	Bill Size and Shape
P. i. sequestratus	inornatus	grayer, more leaden	slightly smaller	smaller
P. i. inornatus[b]	ridgwayi	darker, less ashy	smaller (esp. shorter)	much shorter
	affabilis	less deeply leaden (esp. wing and tail)		very stout and blunt
P. i. transpositus	inornatus	slightly grayer	slightly larger	much heavier
	affabilis	browner, less leaden gray; bill and feet less blackish; retrices and remiges brownish vs. plumbeous		
P. i. mohavensis	transpositus	less olivaceous or brownish; whiter ventrally; flanks and undertail coverts Smoke Gray		longer, on average, than other coastal races
	ridgwayi	much darker, browner gray		shorter
	zaleptus			shorter
P. i. kernensis	inornatus	grayer, less brownish dorsally; flanks and underparts slightly less buffy; more clearly whitish		
P. i. affabilis	transpositus	less olivaceous dorsally, paler gray ventrally		less massive
	zaleptus	less gray dorsally		
	inornatus	more plumbeous, no brown; dark, leaden-colored; bill and feet black		shorter, less massive

Form	Compared to	Plumage characters			
P. i. cineraceus	"typical inornatus"	equally light below, but upper parts ashy gray vs. olive-brown			
	ridgwayi	more gray above; bill black vs. horn-colored			
P. i. zaleptus	sequestratus	much paler, more grayish	larger		
	ridgwayi[c]	more grayish above, with no brownish tinge; paler above; ligher, more grayish below, with little or no buffy wash			
	ridgwayi[d]	darker above and below, with a distinct olive tone above; back is Dark Olive-Gray rather than Deep Olive-Gray			larger, broader (than griseus [= ridgwayi] plus any other form)
	plumbescens	back is Dark Olive-Gray rather than Deep Mouse Gray			
P. i. ridgwayi[e]	inornatus	decidedly grayer; above brownish-gray, beneath paler grayish (versus inornatus = grayish olive-brown, beneath grayish- white)		rather larger	
P. i. plumbescens	ridgwayi	darker, more leaden	tail shorter		smaller (esp. shorter)

[a] Subspecies to which a given form is compared. For example, P. i. sequestratus is: grayer and more leaden, slightly smaller, and has a smaller bill than P. i. inornatus; darker/less ashy and smaller than P. i. ridgwayi; and less deeply leaden, with a much shorter bill, in comparison with P. i. affabilis. See Appendix 2 for further discussion of subspecific taxonomy.

[b] Original description of species (Parus inornatus).

[c] Comparison between zaleptus and ridgwayi based on type description of zaleptus (Oberholser 1932).

[d] Comparison between zaleptus and ridgwayi based on description of zaleptus by Linsdale (1938a).

[e] Originally described as P. i. griseus by Ridgway (1882); renamed as ridgwayi by Richmond (1902).

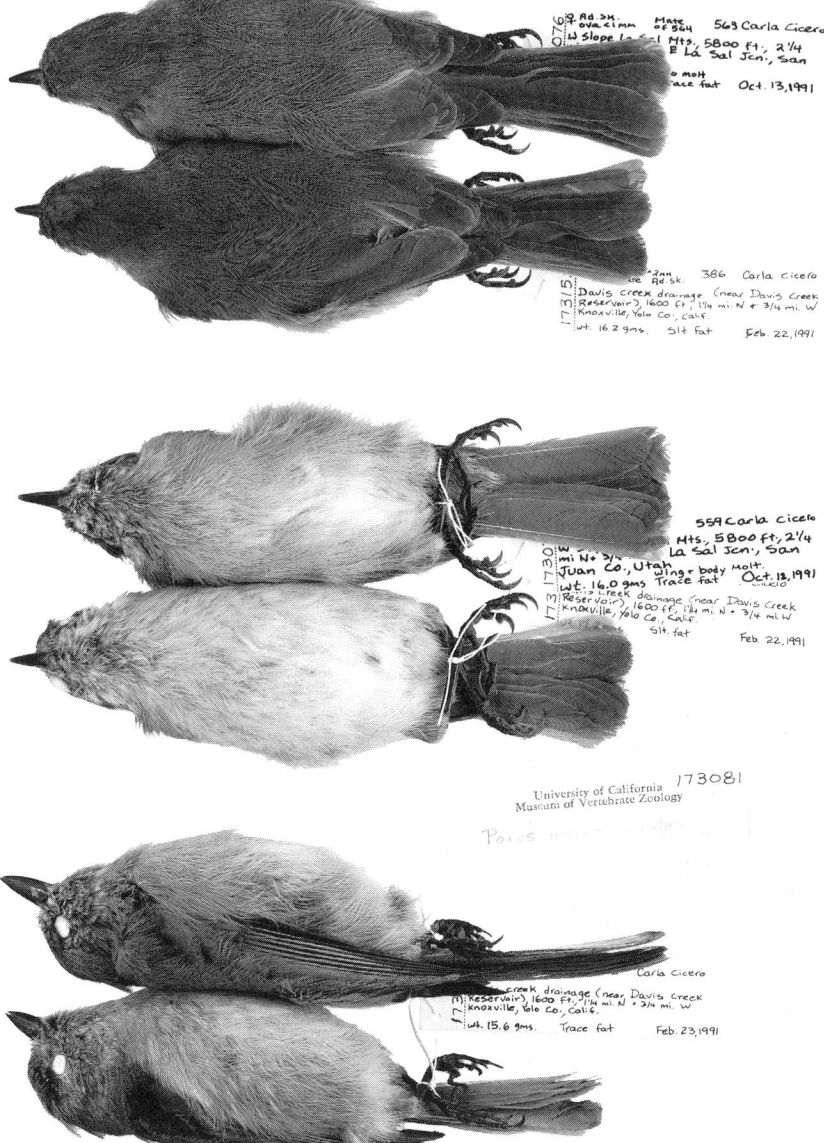

FIGURE 2. Photograph of study skins of titmice, showing the relatively subtle phenotypic differences among populations. In each pair, the bird on the left is from Yolo County, California, and the bird on the right is from San Juan County, Utah. Note the noticeably larger bill, larger overall size, and paler plumage of the Utah specimens.

OBJECTIVES OF THIS STUDY

I used modern systematic techniques to investigate phenetic, genetic, and vocal variation in *Parus inornatus* across its geographic distribution. Earlier workers named subspecies on the basis of morphologic measurements and/or qualitative comparisons of plumage color. These studies were often characterized by small sample sizes and lacked statistical treatment. With recent advances in molecular systematics, additional tools are available to study genetic variability and its geographic concordance with external characters (see Avise 1994). Such data also provide a means for indirectly estimating levels of gene flow among populations. Analyses of mitochondrial DNA (mtDNA) variation have proven to be particularly useful for elucidating phylogenetic relationships at or within the species level because of the relatively rapid rate of animal mtDNA evolution coupled with its typically uniparental mode of inheritance (A.C. Wilson et al. 1985; Avise et al. 1987; Moritz et al. 1987; Shields and Helm-Bychowski 1988; Avise 1994). Although mtDNA sequencing methods provide the most detailed level of resolution for examining evolutionary questions, few workers have used this approach to assess genic variation within bird species (e.g., N.K. Johnson and Cicero 1991; Edwards and Wilson 1990; Birt-Friesen et al. 1992). Rather, most intraspecific studies of mtDNA variability have employed the faster method of restriction fragment analysis (e.g., Shields and Wilson 1987a; Ball et al. 1988; Tegelström et al. 1990; Zink et al. 1991; Ball and Avise 1992; Hare and Shields 1992), and studies of mtDNA sequence variation across the geographic distributions of avian taxa are essentially lacking.

While this study resulted in a taxonomic revision of the *Parus inornatus* complex, the primary goal was not nomenclatural. Rather, the main objective was to assess patterns of variation in several suites of characters at the level of *populations* rather than subspecies. The basic null hypothesis was that populations are invariant in all of the characters examined. Several questions were addressed: (1) What are the patterns of variation in morphology, color, allozymes, mtDNA, and voice within and among populations in different parts of the range? (2) What is the level of geographic concordance exhibited by different suites of characters? (3) Do the patterns of differentiation correspond to environmental gradients and/or ecological associations? (4) What is the degree of morphologic and/or genetic change across isolating barriers? (5) What evolutionary processes (e.g., gene flow, post-Pleistocene disjunction) may be invoked to explain inter-population differentiation? And (6) Do the observed patterns of variation correspond to currently recognized taxonomic limits?

MATERIALS AND METHODS

A major concern in studies of geographic variation in birds is the availability of breeding material. Many previous workers have focused on taxa that are migratory or contain at least some migratory populations, where it is essential to discriminate between breeding and non-breeding specimens. Such discrimination is especially difficult given the lack of reproductive data associated with many museum skins, and the uncertainty of using dates of capture as an indication of breeding status. Another limitation with such studies is the relatively brief season for fieldwork. The resident status of *P. inornatus* confers a major advantage for the systematic study of variability in this species. In addition to enabling year-round fieldwork, both breeding and non-breeding specimens could be used from museum collections. Thus, the number of specimens examined in this study exceeds that of most other treatises on intraspecific variation in birds. The non-migratory habit of *P. inornatus* also has important implications for the interpretation of patterns of genetic and phenotypic variation.

In order to be as objective as possible, previously established subspecies boundaries were ignored for the purposes of data collection and analysis. However, the resulting patterns of variation are presented and discussed in the context of currently recognized subspecies in order to test their validity.

FIELDWORK

The field portion of this study began in the spring of 1988 and continues to the present, with the most concentrated work occurring between 1989 and 1991 (total of 166 days in the field). Field seasons extended from late February through June and from September through October or mid-November. Locality data from museum skins were helpful in identifying sites with adequate numbers of titmice; additional localities were discovered by exploration. Field effort emphasized the collection of 10 to 20 specimens per population from sites distributed throughout the geographic and ecologic range of the species in the western United States; the cost and difficulty of obtaining permits for Mexico precluded work in that country. Permit limitations, in conjunction with population size, determined the number of specimens taken from each deme. Specimens were collected by shotgun according to standards accepted by the American Ornithologists' Union (1988) and by the U.C. Berkeley Animal Care and Use Committee. Although the main objective of fieldwork was to obtain tissue samples for genetic analysis, specimens also were collected to provide supplemental material for the study of variation in morphometric characters and plumage coloration.

Body weight (gms) was taken from freshly killed specimens using a Pesola spring scale. Tissue was excised within a few hours after collection and temporarily preserved in liquid nitrogen (-196°C) until transport to the Museum of Vertebrate Zoology, Berkeley, for permanent storage at -76°C. Specimens were prepared either as standard study skins or as skin-skeletons (N.K. Johnson et al. 1984). Detailed notes on reproductive condition, pairing and family relationships (where possible), skull ossification, molt, habitat, and behavior were written daily.

Tape recordings of singing males were obtained from a few populations for a preliminary assessment of vocal variation within and among populations. Recordings were made using a Sony TCM-5000EV professional recorder with a Sennheiser ME-88 shotgun microphone and a BA-3 nature amplifier. Some recordings also were obtained by Ned K. Johnson during his fieldwork in the western United States. A more comprehensive analysis of vocal variation in western titmouse populations will be presented in another study.

SPECIMENS EXAMINED

Specimens of *Parus inornatus* are well represented in museums that house collections of western North American birds. To thoroughly assess the geographic distribution of this species, and to obtain adequate sample sizes for the morphometric and color analyses, study skins were borrowed from every major and minor collection of titmice in the United States. A total of 3362 specimens was examined (Table 2) from the following institutions: American Museum of Natural History, New York; Monte L. Bean Life Science Museum, Brigham Young University; Bell Museum of Natural History, University of Minnesota; California Academy of Sciences, San Francisco; California State University, Long Beach; California State University, Sacramento; Carnegie Museum of Natural History, Pittsburgh; Cleveland Museum of Natural History; Cornell University; Delaware Museum of Natural History; Denver Museum of Natural History; Field Museum of Natural History, Chicago; Idaho Museum of Natural History; Los Angeles County Museum of Natural History; Museum of Comparative Zoology, Harvard University; Museum of Natural History, Kansas University; Museum of Natural Science, Louisiana State University; Museum of Northern Arizona, Flagstaff; Museum of Southwestern Biology, University of New Mexico; Museum of Vertebrate Zoology, Berkeley

TABLE 2. Sample areas and specimens examined.

Sample Area[a]	Morphometrics Males	Females	Reflectance Spectro.	Allozyme Electro.	mtDNA Seq.	Total
1. Medford	34	21	17	15	--	73
2. Siskiyou	21	16	12	31	4	52
3. Mt. Dome	15	8	--	25	2	25
4. North Sacramento Valley	36	29	3	16	--	76
5. Clear Lake	41	18	2	17	2	73
6. Oroville	24	22	13	14	2	74
7. San Francisco Bay Area-North	33	22	9	--	--	71
8. San Francisco Bay Area-South	105	82	38	--	--	232
9. Stockton	7	2	1	--	--	13
10. Monterey	56	42	20	16	2	120
11. San Benito	30	22	7	--	--	70
12. San Luis Obispo	27	15	9	16	2	43
13. W Slope Sierra Nevada-North	26	11	4	--	--	38
14. W Slope Sierra Nevada-Central	16	10	1	--	--	34
15. W Slope Sierra Nevada-South	18	11	--	15	2	35
16. Kern	44	21	15	15	2	98
17. Santa Barbara	14	10	7	--	--	34
18. Tejon	25	20	15	17	2	51
19. Ventura	23	14	8	--	--	38
20. Newhall	16	11	5	--	--	28
21. Los Angeles	64	38	29	--	--	112
22. San Bernardino Mts.	27	14	12	--	--	50
23. San Jacinto Mts.	11	8	2	--	--	28
24. Joshua Tree	37	22	8	10	2	64
25. Orange	10	5	1	--	--	18
26. San Diego-North	44	37	16	--	--	85
27. San Diego-South	63	38	4	17	2	113
28. Sierra Juarez	16	14	11	--	--	34
29. Sierra San Pedro Mártir	18	15	11	--	--	37
30. Sierra de la Laguna	84	63	25	--	--	177

TABLE 2 (continued)

Sample Area[a]	Morphometrics		Reflectance Spectro.	Allozyme Electro.	mtDNA Seq.	Total
	Males	Females				
31. Warner	18	12	--	16	2	36
32. Tule Lake	11	7	--	13	2	19
33. NE California	29	17	12	18	2	61
34. Reno	22	17	1	13	--	43
35. Benton	11	11	--	15	--	27
36. White-Inyo Mts.	33	27	12	15	2	70
37. Eastern Mojave	48	37	3	15	2	96
38. Southern Nevada	4	3	1	--	--	16
39. SE Nevada	15	9	--	18	2	35
40. Ruby Mts.	15	7	--	15	2	23
41. Southern Idaho	16	16	8	--	--	38
42. N-Central Utah	25	17	10	19	2	61
43. Central Utah	8	5	4	--	--	15
44. La Sal	17	20	1	16	--	42
45. New Castle	18	18	7	16	2	48
46. Pueblo	5	7	2	--	--	14
47. SW Plains	22	10	5	--	--	38
48. N-Central New Mexico	20	10	5	20	1	51
49. Zuni Mts.	9	7	--	17	2	17
50. Four Corners	7	5	1	--	--	19
51. SW Utah	10	7	2	--	--	22
52. Kaibab	13	9	3	15	2	24
53. San Francisco Mts.	19	22	2	17	1	51
54. Central Arizona	16	8	7	--	--	32
55. Eastern Arizona	24	12	7	--	--	37
56. Reserve	67	51	5	15	2	123
57. Silver City	18	7	7	--	--	29
58. Chiricahua Mts.	17	15	5	12	2	34
59. S-Central New Mexico	31	17	2	16	--	52
60. Guadalupe Mts.	4	5	5	--	--	10
Totals[b]	1557	1076	422	525[c]	52[c,d]	3179

[a] See Figs. 3-4. Exact localities are given in Appendix A.

[b] An additional 130 specimens were examined from localities plotted in Figs. 3-4 but excluded from sample areas; 53 specimens were examined from localities not plotted in Figs. 3-4. See Appendix A for detail.

[c] Excludes outgroup specimens of *P. bicolor* and *P. gambeli* (localities given in Appendix A).

[d] An additional sequence was obtained from 1 individual from Fairview Peak, Churchill Co., Nevada (outside of sample areas).

(approximately one-third of all specimens examined); Museum of Zoology, University of Michigan; National Museum of NaturalHistory, Washington, D.C.; Nevada State Museum; Oklahoma Museum of Natural History; Oregon State Museum of Anthropology; Philadelphia Academy of Natural Sciences; San Bernardino County Museum of Natural History; San Diego Museum of Natural History; San Francisco State University; Santa Barbara Museum of Natural History; Texas Cooperative Wildlife Collection, Texas A&M University; University of Arizona; University of California, Los Angeles; University of Colorado, Boulder; University of Nevada, Reno; University of Puget Sound, Tacoma; Utah Museum of Natural History; Utah State University, Logan; and Western Foundation of Vertebrate Zoology, Los Angeles.

MUSEUM AND LABORATORY METHODS

Measurement of Morphologic Characters

Eight linear measurements were taken from study skins using Fowler Ultra-cal II digital calipers interfaced with an IBM portable computer. Only birds of known sex were measured. Age data were recorded from specimen tags where available and categorized into three groups: immature, adult, and unknown; over 50% of the specimens lacked age information. Juvenile birds with fluffy plumage were ignored. Measurements were taken to the nearest 0.1 millimeter as follows: (a) wing length: chord length from the bend of the wing to the tip of the longest primary feather; (b) tail length: distance from the basal insertion of the central pair of rectrices to the tip of the longest rectrix; (c) tarsus length: the diagonal of the tarsometatarsus, from the posterior juncture of the tibia and metatarsus to the edge of the lowest undivided scute at the anterior junction of the metatarsus and base of the middle toe (see diagram in Baldwin et al. 1931:107); (d) length of middle toe: the diagonal of the middle toe (without claw), from the anterior edge of the lowest undivided scute to the tip of the toe pad on the ventral surface of the toe; (e) hallux length: distance from the tip of the toe pad on the ventral surface of the first digit to the ventral base of the pad at the juncture with digit II; (f) bill length: distance from the anterior edge of the nostril to the tip of the upper mandible; (g) bill width: distance from one tomium to the other, taken where a plane drawn at right angles to the bill passes through the anterior margin of the nostril; (h) bill depth: distance from the culmen to the lower edges of the rami, taken along the same plane as bill width.

Weight (gms) was recorded from tags on museum skins and transformed to cube roots for analysis. Because many specimens lacked such data, the analysis of variation in body weight is more limited than that for other characters. The reliability of data on body weight may be confounded by seasonal, and not necessarily geographic, variation in reproductive condition and fat deposition. Although this generally poses a greater problem for migratory bird species, weights were ignored for female titmice in advanced stages of reproduction (e.g., egg-laying).

Quantitative Appraisal of Plumage Coloration

Numerous methods of color analysis are available, including both qualitative and quantitative measures (see Bowers 1956). Most studies of color variation in birds have used qualitative approaches that involve one of the following: visual comparison among series of museum skins (e.g., Behle 1956; Storer and Zimmerman 1959; Banks 1988); scoring of plumage characters based on discrete categories (e.g., Temple 1972; Gill 1973; Atwood 1988; Hill 1992); or matching of plumage colors to standards such as the Munsell color system (e.g., Banks 1986; Bell 1992). Despite the few quantitative assessments of color variation in plumage (e.g., N.K. Johnson and Brush 1972; N.K. Johnson 1980; Dyck 1992), such methods offer a more objective appraisal of variability. In this study, I used a Bausch and Lomb Spectronic 505 recording spectrophotometer to analyze variation in color of the central back, breast, and belly plumage of *P. inornatus*. The analysis was limited to specimens with unworn plumage collected primarily between September and March. Although specimens collected up to 90 years ago were examined, few individuals showed visible signs of foxing; those that did were excluded from measurement. Non-uniform surfaces (e.g., ruffled or soiled feathers, open belly areas) also were not measured. Unfortunately, the instrument broke irreparably in the middle of the study and thus many potentially useful specimens could not be analyzed.

The spectrophotometer was equipped with a visible reflectance attachment, and samples were measured through a 22 mm port. Flatness of the 100% line was maintained within limits of 1% peak-to-peak; flatness of the 0% line was calibrated to within 0.5%. Curves of percentage diffuse spectral reflectance between 400 and 700 mμ were used to calculate trichromatic coefficients (x, y, z) of Illuminant C by the 10 selected ordinate method of Hardy (1936:49-51). From these coefficients, dominant wavelength (hue), relative brightness (value), and excitation purity (chroma) were computed according to Table 6 in Judd (1933:371); computations were made on a programmable Texas Instruments calculator using a program written by O. P. Pearson of the MVZ.

Allozyme Electrophoresis

Because specific enzymes are associated with certain tissue types (see Harris and Hopkinson 1976), the most efficient method of surveying a broad range of loci is to combine tissues for analysis. Accordingly, small amounts of tissue (heart, breast muscle, liver, and kidney) frozen in the field were minced together to form a homogenized mass for the preparation of protein extracts. Extracts were prepared by combining this tissue mixture with an equal volume of de-ionized water and centrifuging it for 30 min at 15,000 rpm (-5°C). The aqueous samples were stored at -76°C for later analysis.

Allozyme variation was examined for 525 individuals of *Parus inornatus* collected from 32 populations throughout the geographic distribution of the species (Table 2); 5 individuals of *P. gambeli* and 3 individuals of *P. bicolor* were analyzed as outgroups.

TABLE 3. Electrophoretic conditions[a] used in the analysis of allozyme variation among 32 populations of the *Parus inornatus* complex and 2 outgroups (*P. bicolor, P. gambeli*).

Gel Buffer[b]	Electrode Buffer[b]	Volts	Time[a,c]	Loci[a,b,d,e]
Lithium (LiOH) A+B, pH 8.2	LiOH A, pH 8.1	300	3 hours	NP; EST-D; LA-1; LGG; GDA; AB-1,2,3,4
Tris Citrate II, pH 8.0 (1:30 dilution of electrode buffer)	Tris Citrate II, pH 8.0	130	4 hours	ADH; αGPD; SDH; MDH-1,2; ICD-1,2; GLUD; GOT-1,2; GPT; PGM; EAP; ADA; MPI
Poulik, pH 8.7	Borate, pH 8.2	250	3 hours	LDH-1,2; SOD-1,2; CK-1,2,3
Tris Maleic EDTA, pH 7.4 (1:10 dilution of electrode buffer)	Tris Maleic EDTA, pH 7.4	100	6 hours	6PGD
PGI Phosphate, pH 6.7 (1:20 dilution of electrode buffer)	PGI Phosphate, pH 6.7	130	6 hours	GPI

[a] Electrophoretic conditions are similar to those described by N.K. Johnson et al. (1984), with slight modifications. Heart, liver, kidney, and breast muscle tissue were ground and analyzed together.

[b] Recipes for gel buffers, electrode buffers, and protein stains are described by Selander et al. (1971) and Harris and Hopkinson (1976).

[c] Comparison gels occasionally were run longer to resolve differences in band mobilities.

[d] Abbreviations for loci are those used by Harris and Hopkinson (1976). Enzyme commission (E.C.) numbers for loci are as follows (International Union of Biochemistry 1984): ADH (1.1.1.1), αGPD (1.1.1.8), SDH (1.1.1.14), LDH (1.1.1.27), MDH (1.1.1.37), 1CD (1.1.1.42), 6PGD (1.1.1.44), GLUD (1.4.1.3), SOD (1.15.1.1), NP (2.4.2.1), GOT (2.6.1.1), GPT (2.6.1.2), CK (2.7.3.2), PGM (2.7.5.1), EST-D (3.1.1.1), EAP (3.1.3.2), LA-1 (3.4.11.1), LGG (3.4.11.4), GDA (3.5.4.3), ADA (3.5.4.4), MPI (5.3.1.8), GPI (5.3.1.9).

[e] Additional loci dropped because of unreliable scoring include: ME (1.1.1.40), Gd (1.1.1.49), GSR (1.6.4.2), PGM-2 (2.7.5.1), ACP (3.1.3.2), PAP (3.4.11), ALD (4.1.2.13), and ACON (4.2.1.3).

Forty-one presumptive genetic loci were run initially to screen for allozyme variability. Eight of these were dropped because of unreliable scoring, and thus a total of 33 loci was examined in this study (Table 3). Techniques generally followed the standard procedures described by Selander et al. (1971) and N.K. Johnson et al. (1984).

Gels were scored conservatively in all cases, and ambiguous samples were rerun as many times as necessary to resolve differences in mobility. Alleles (electromorphs) at each locus were designated alphabetically in order of decreasing mobility from the origin. For proteins with multiple isozymes, the most anodal locus was identified as 1; progressively higher numbers were assigned to more cathodal loci. Gels with ultraviolet stains were photographed for future reference. Other gels were fixed with a solution of methanol/acetic acid and then either dried and mounted on 5 x 7 index cards (for agar overlay stains) or wrapped in plastic and refrigerated at -4°C.

Amplification and Sequencing of mtDNA

Using the polymerase chain reaction (PCR) to amplify fragments of DNA (Kocher and White 1989; White et al. 1989), systematists have focused their attention recently on obtaining direct nucleotide sequences in order to resolve phylogenetic relationships. Early studies of mtDNA sequencing in vertebrates were limited primarily to the cytochrome *b* gene (e.g., Kocher et al. 1989). Although cytochrome *b* continues to dominate most sequencing studies, workers have begun to explore other mitochondrial genes (Thomas et al. 1990; Quinn 1992) as well as nuclear genes (M.F. Smith et al. 1992). The utility of cytochrome *b* for assessing genetic variability depends on several factors, including the type of organism being studied and the taxonomic level of concern (Graybeal 1993). Nonetheless, cytochrome *b* continues to provide valuable information on phylogenetic relationships within and between species (e.g., N.K. Johnson and Cicero 1991; M.F. Smith and Patton 1991; Patton and Smith 1992; Cicero and Johnson 1995) as well as among more distant lineages (Edwards et al. 1991).

I amplified and sequenced 300 base pairs (bp) of cytochrome *b* from 53 individuals of *P. inornatus* sampled from 26 (81%) of the 32 populations analyzed allozymically (Table 2). An additional individual of *P. inornatus*, collected from an isolated population in west-central Nevada, also was sequenced. One specimen of *P. bicolor* and two *P. gambeli* were sequenced as outgroups. DNA was obtained from 10-20 mg of frozen liver tissue by digesting the tissue overnight in 500 μl lysis buffer (50 mM Tris HCL, pH 8.0; 50 mM ethylenediamine tetraacetate [EDTA], pH 8.0; 1% sodium dodecyl sulfate; 100 mM NaCl; 1% 2-mercaptoethanol) and 11 μl proteinase K. Tubes were incubated at 55°C and gently rocked until all of the tissue was dissolved. RNase A (5.5 μl) was added to each sample one hour before the end of incubation. Whole-genomic DNA was purified either by (1) extracting once with phenol (pH 8.0), once with phenol:SEVAG (1:1), and once with SEVAG (24 chloroform:1 isoamyl alcohol), or (2) extracting once with 5M sodium chloride. Cold, absolute ethanol concentrated the DNA into "ropes." Pellets of DNA obtained by microcentrifuging were washed twice with 70% ethanol and re-suspended in 100 μl 1X Tris EDTA (pH 8.0) at 55°C. Two controls without tissue were included in each set of extractions to test for laboratory contamination. The quality and molecular weight of DNA were assessed by running the extracts on a 1% HGT

TABLE 4.   Locations and sequences of cytochrome *b* primers used for amplification and sequencing.

Primer Location[a]	Primer Sequence[b]
L14987	5'-CCATCCAACATCTC[A/T]GC[A/T]TGATG-3'
H15304	5'-GTAGCACCTCAGAA[G/T/C]GATATTTG-3'
H15916	5'-ATGAAGGGATGTTCTACTGGTTG-3'

[a] Letters refer to light (L) and heavy (H) strands of mtDNA; numbers correspond to 3' end of primer in chicken (*Gallus*) sequence (Desjardins and Morais 1990).

[b] Degenerate sites are indicated by brackets ([]).

agarose minigel in 1X TBE buffer, and by staining with ethidium bromide (10 μg per ml). Extracts were stored in 1X TE buffer (pH 8.0) at 4°C or -20°C.

Primers L14987 and H15916 (Table 4) were used to amplify double-stranded DNA fragments of approximately 924 bp.   Each double-stranded PCR (dsPCR) reaction contained 12.5 μl of the target DNA (diluted by varying amounts in 1X low TE or water) and 12.5 μl of a mixture with final concentrations of 1X TAQ buffer (Cetus Corp.; 10 mM Tris pH 8.3, 1.5 mM $MgCl_2$, 50 mM KCl, 0.01% gelatin), 0.75 mM dNTP mix, 1 μM of each primer, 0.625 units of *Thermus aquaticus* (Taq) polymerase, and double-distilled water.   Each cycle of dsPCR involved denaturation at 92°C for 1 min, annealing at 50°C for 1 min, and extension at 72°C for 1.5 min (30 cycles).   Products were resolved on 2% Nusieve agarose gels in 1X TA buffer and stained with 10 μg/ml ethidium bromide.   Plugs of amplified double-stranded DNA were excised from agarose gels, diluted in 250 μl 1X low TE, and melted in a 65°C water bath.

The unbalanced primer method (Gyllenstein and Ehrlich 1988) was used to obtain single-stranded DNA from the melted agarose plugs of dsPCR products.   Approximately 300 bp of single-stranded template was amplified using primers L14987 and H15304 (Table 4).   Single-stranded PCR (ssPCR) reactions were performed in 50 μl volumes containing 10 μl of the diluted dsPCR plugs, 15 μl of double-distilled water, and 25 μl of PCR reagents.   The PCR solution was similar to that used in dsPCR except for the type of buffer (New England Biological Lab; 67 mM Tris pH 8.8, 2 mM $MgCl_2$, 16.7 mM ammonium sulfate, 10 mM β-mercaptoethanol) and the relative amount of each primer (50:1 ratio, with final concentrations of 10 μl to 0.2 μl).   Two sets of ssPCR reactions were performed on each dsPCR product to obtain an excess of one strand over the other; the 50:1 primer ratio was reversed when amplifying the complimentary strand. Single-stranded reactions were performed for 32 cycles under the same conditions as dsPCR amplification.   Bands were assessed by electrophoresis (3% HGT and 1% Nusieve agarose minigels in 1X TAE buffer) with ethidium bromide staining (10 μg/ml).

Thermal cycling for dsPCR and ssPCR was performed in a Techne PHC-2 programmable heating block (Perkin Elmer-Cetus). All reaction volumes were layered with 1-2 drops of mineral oil to prevent evaporation during heating. Two negative controls comprised of water and the PCR mixture were included in each set of experiments to ensure that the amplified sequences were not contaminated from other sources. Protocols for preparing stock solutions for both DNA extraction and amplification followed Maniatis et al. (1982). Double-stranded and single-stranded products were kept at 4°C until further analysis, or at -20°C for long-term storage.

Single-stranded DNA was cleaned of free nucleotides and excess salts by either (1) subjecting samples to 3-4 cycles of centrifugation dialysis (Centricon 30; Amicon Corp.), or (2) spinning the samples using Quick-Spin^{TM} G-50 Sephadex Columns (Boehringer Mannheim Corp.) with pre-swollen beads. Concentrated DNA purified by centrifugation dialysis was stored in 20-25 μl of double-distilled water (-20°C). Sequencing reactions were performed using 7 μl of clean DNA template and the limiting primer in ssPCR amplification. A commercial kit (Sequenase, US Biochemical Corp.) was used for sequencing according to the Sanger dideoxy chain-termination method (Sanger et al. 1977). The resulting products were loaded onto a 6% polyacrylamide-8.3M urea linear gel (1x TBE buffer), run for 1.5-5 hours at 40-45°C, and autoradiographed.

DATA ANALYSIS

Pooling of Samples

Although point samples are ideal for studying patterns of geographic variation (R.S. Thorpe 1976; Zink and Remsen 1986), most of the titmouse localities were represented by insufficient numbers of skins to permit point sample analysis of size and color data. Consequently, it was necessary to pool localities in order to increase sample sizes. Specimens were pooled into 60 areas (Figs. 3-4 and Table 2) distributed throughout the range of *P. inornatus*. The size and geographic distribution of sample areas was dictated by the availability of specimens, with the greatest representation from California (see Appendix A). Pooled sample areas were as small as possible, taking into account the number of specimens of each sex. As is typical for museum collections, males were much more numerous than females. A minimum sample size of 10 per sex was targeted for the morphometric analysis, which is in line with other studies of geographic variation in birds. However, sample sizes for some areas were smaller because of the limited availability of material, especially for females. Most of the specimens examined morphometrically were collected during the late spring and summer when the plumage is very worn, and thus were not useful for color analysis. Although suitable specimens for spectrophotometry were available from 52 of the 60 sample areas, sample sizes were small for several areas. Therefore, color data were analyzed for a maximum of 45 populations; breast and belly color were analyzed for fewer populations

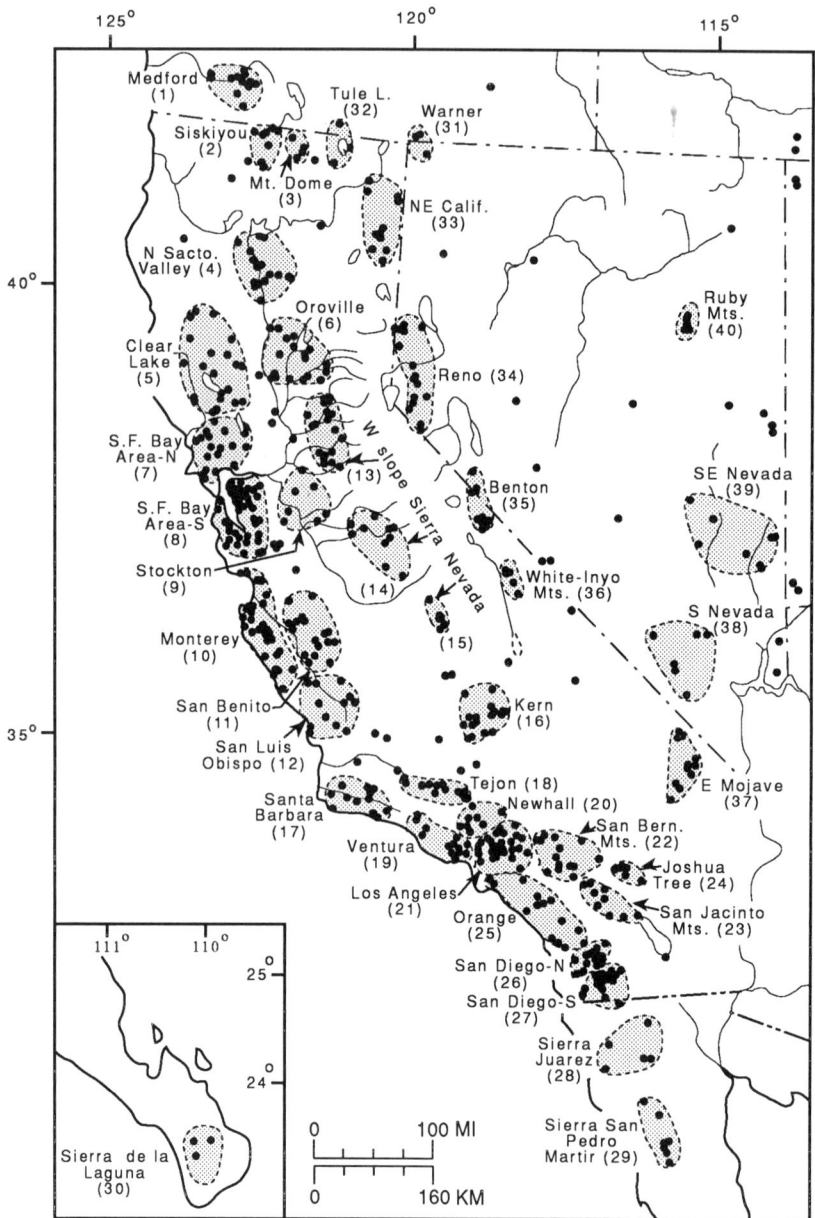

FIGURE 3. Localities (dots) of all specimens of the *P. inornatus* complex examined from Oregon, California, Baja California, and Nevada. Inset shows the isolated population in the Cape District (Sierra de la Laguna Mountains) of southern Baja California. Dashed lines enclose sample areas used for analysis. Sample area names and numbers (in parentheses) are the same as in Table 2. See Appendix A for specific localities.

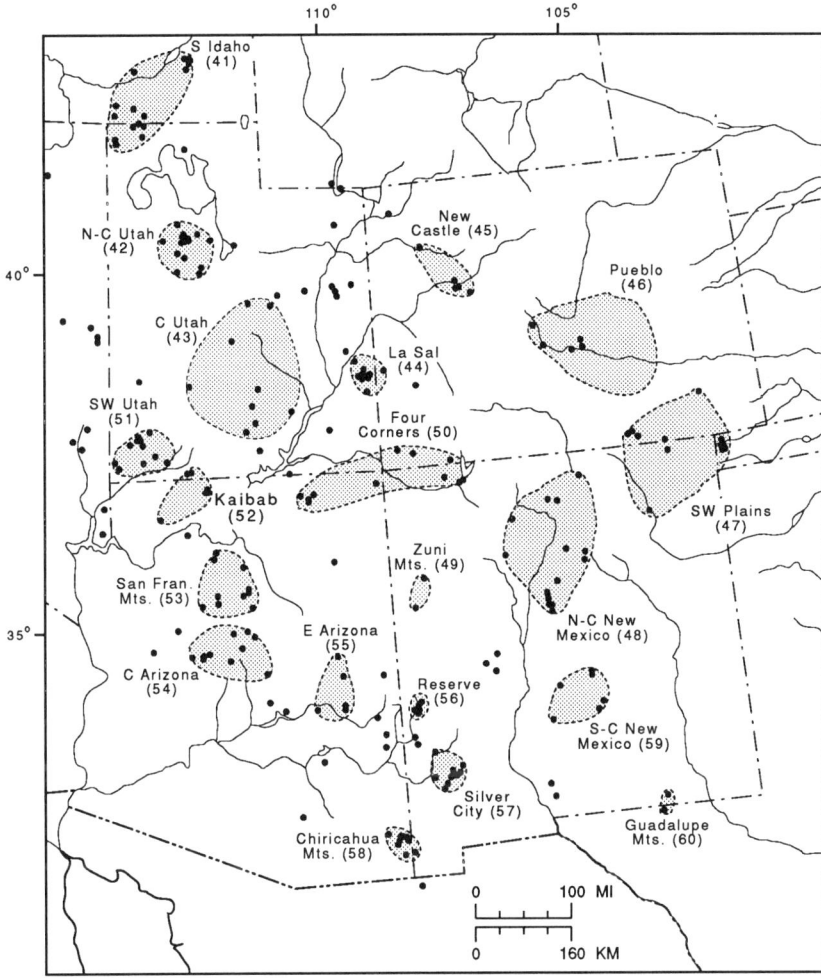

FIGURE 4. Localities (dots) of all specimens of the *P. inornatus* complex examined from southern Idaho to northwestern Texas. Dashed lines enclose sample areas used for analysis. Sample area names and numbers (in parentheses) are the same as in Table 2. See Appendix A for specific localities.

because of missing data resulting from soiled or uneven plumage. Tissue samples were collected from 32 of the 60 sample areas.

Natural features (e.g., habitat breaks) that might act as barriers to gene flow also were considered in the determination of sample areas. Uncertain localities that could not be mapped were excluded from sample areas unless the sample area covered the entire range of possible sites; for example, sample area 46, which encompasses Pueblo County, Colorado, includes specimens taken from that county from unknown specific localities. The total number of sample areas (60) was limited by the computer routine used to run the sums-of-squares simultaneous test procedure (see below).

Analysis of Phenotypic Variation

Analysis of non-geographic variation. Multivariate analysis-of-variance (MANOVA) was used to assess the effects of age and sex, independent of sample area, on size variation in *P. inornatus*. MANOVA also was used to test the effect of sex on patterns of color variation. Coefficients of variation were calculated for each character to examine intra-population variability in size and color.

Biometric analysis of geographic variation in size and color. I employed both univariate and multivariate approaches in the analysis of morphometric and color variation. Basic descriptive statistics were computed using the Frequencies routine in SPSS^X (SPSS Inc.1986). In order to expose serious redundancies among different size and color characters (see R.S. Thorpe 1976), pairwise coefficients of correlation were calculated for each data set and then subjected to clustering. Simple linear regression was used to assess correlations between males and females in univariate size patterns and to examine the relationship between different color variables.

Geographic variation in individual size and color characters was analyzed by the sums-of-squares simultaneous test procedure (SS-STP; Gabriel 1964; Gabriel and Sokal 1969), using the program from Sokal and Rohlf (1969) as modified into the algorithm PAIRS by Bruce Krogman and Alan Stangenberger. This method statistically tests differences in population means based on the magnitude of each mean as well as associated variances and sample sizes (see Power 1970). Means are ranked in order of magnitude, with vertical lines denoting maximally homogeneous subsets of means for individual characters based on an *a posteriori* analysis-of-variance of all samples combined (*F*-statistic non-significant at $P \leq 0.05$). A criticism of this approach has been that patterns of variation may be difficult to interpret because localities are arrayed by character means regardless of geography (Barrowclough 1992). In order to circumvent this problem, workers have displayed the results of SS-STP analyses in the form of pie diagrams superimposed onto maps of geographic samples (e.g., Power 1970; N.K. Johnson 1980; Atwood 1988; N.K. Johnson and Marten 1992). I followed the method of scaling used by N.K. Johnson and Marten (1992:5), in which the total range of means for each character was converted to 100% and each mean was scaled according to its relative magnitude. Thus, a mean that was intermediate between the lowest and highest value received a scale of 50% on the pie diagram. Geographic differences in the relative scale of each pie can be used to assess patterns of clinal variation, zones of abrupt character change, and similarities (i.e., convergence) among geographically distant populations.

Multivariate patterns of size and color variation were analyzed using an array of phenetic techniques that provide different but complimentary levels of resolution. Separate matrices of population means for size and color characters were standardized by variance prior to multivariate analysis. NTSYS-pc (Rohlf 1988) was then applied to these data to cluster and ordinate populations. These two broad categories of methods complement each other in that clustering more accurately depicts close relationships, whereas ordination more faithfully portrays larger groupings (Sneath and Sokal 1973). A disadvantage of clustering for intraspecific studies, however, is that populations of intermediate character (e.g., hybrids or intergrades) are not revealed.

Populations were clustered by Euclidean distances analyzed with the unweighted pair group method (UPGMA). Ordination techniques included principal-components analysis (PCA) and non-metric multidimensional scaling (MDS, using MDSCALE). Unlike PCA, in which the main objective is to maximize the percentage of variance explained by the vectors, non-metric MDS attempts to project points such that inter-point distances in the ordination plot most closely reflect the original distance matrix (Rohlf 1988). Accordingly, MDS potentially improves the principal-component plot by iteratively comparing inter-point distances on the plot with original Euclidean distances until minimum "stress" is achieved (Rohlf 1988). Input into the MDS analysis included the Euclidean distance matrix and an initial configuration based on PCA. Minimum spanning trees (prim networks) also were produced from the Euclidean distance matrix to link populations and to reveal distortions in the ordination plots (Sneath and Sokal 1973).

Analysis of Genetic Variation

Biometric analysis of allozyme variation. Raw gene frequency data were analyzed using the program BIOSYS-1 (Swofford and Selander 1981) to compute the following statistics: allelic frequencies, observed and expected heterozygosities, percentage of loci polymorphic, mean number of alleles per locus, X^2 tests for departure from Hardy-Weinberg equilibrium, genetic distances, and S. Wright's (1965) F-statistics using the modifications of S. Wright (1978) for small sample size and Nei (1975) for multiple alleles. Average heterozygosities were determined by direct count ($H_{obs.}$) and from estimates ($H_{exp.}$) based on Hardy-Weinberg expectations (Nei 1975). Deviations from Hardy-Weinberg were tested using the chi-square goodness-of-fit test with Levene's (1949) correction for small sample sizes; because of problems with this test when expected frequencies are low (Sokal and Rohlf 1969), genotypes were pooled and retested for loci with more than two alleles (see Swofford and Selander 1981:17). Allelic frequencies were transformed to genetic distances according to the methods of Nei (1978), J.S. Rogers (1972), and Cavalli-Sforza and Edwards (1967). Each distance measure has its own set of properties and assumptions (see S. Wright 1978; Farris 1981; Nei et al. 1983; Hillis 1984; Swofford and Olson 1990). A major advantage of Cavalli-Sforza and Edwards' (1967) estimates is that they are not affected by differences in intra-population heterozygosity (Swofford and Olsen 1990). Both the chord and arc distances provided by Cavalli-Sforza and Edwards (1967) assume that all genetic change is attributed to random drift, independent of initial gene frequencies.

Controversies surrounding the use of genetic distance measures also apply to techniques for constructing trees based on distance data (see Bledsoe 1987; Swofford and Olson 1990:436-451; Avise 1994:109-120). Two major categories of algorithms are available: phenetic methods such as UPGMA clustering, which assume constant rates of evolutionary change according to a molecular-clock hypothesis (J.P. Thorpe 1982); and phylogenetic approaches that do not assume uniform rates of gene substitution. While the most commonly used method in phylogenetic studies of allozyme variation is the distance Wagner tree (Farris 1972), this technique has several disadvantages (see Prager and Wilson 1978; Swofford 1981; Nei et al. 1983). Another approach that has been used relatively recently with genetic distance data is the neighbor-joining method (Saitou and Nei 1987). This algorithm operates under the principle of minimum evolution, building a tree by minimizing the sum of branch lengths between pairs of "neighbors" (operational taxonomic units) at each stage of clustering (see Saitou and Nei 1987 for detailed description). Although the neighbor-joining (NJ) method uses the same means of estimating branch lengths as the Fitch-Margoliash (1967) procedure (Saitou and Nei 1987), it has a major advantage in terms of computational speed (Felsenstein 1991). During computer simulations, the neighbor-joining method outperformed other algorithms in recovering the correct tree topology (Saitou and Nei 1987).

I applied the UPGMA and neighbor-joining procedures to Cavalli-Sforza and Edwards (1967) chord distances to assess relationships among populations of *Parus inornatus*. Trees were produced using the NEIGHBOR program in PHYLIP (Felsenstein 1991), with *P. bicolor* as the outgroup species; *P. gambeli* was excluded because of the relatively large genetic distances separating this species from populations of *P. inornatus*, thereby obscuring relationships within the ingroup. Unlike BIOSYS-1, PHYLIP divides the chord distance by the degrees of freedom (sum of the number of alleles minus one) and squares the entire value; this quantity (D^2) is expected to rise linearly with the amount of genetic drift (Felsenstein 1991). The distance data were analyzed by using the input order of populations and by randomizing the sequence of populations. Because PHYLIP does not provide a cophenetic correlation coefficient for UPGMA trees, the value had to be computed by reanalyzing the data using NTSYS-pc (Rohlf 1988). This was accomplished by re-clustering the data based on the genetic distance matrix calculated from PHYLIP and comparing the genetic distances with a cophenetic matrix produced from the UPGMA tree. Goodness-of-fit statistics were not available for the neighbor-joining tree.

Because hierarchical techniques cannot depict patterns of reticulate evolution, hybridization, and/or clinal change, recent studies have applied ordination techniques to genetic distance data in order to examine geographic variation within species (Van Wagner and Baker 1986; Barrowclough and Johnson 1988; Lessa 1990). Using the Cvalli-Sforza and Edward (1967) chord distance matrix from PHYLIP, I further examined allozymic variation in *P. inornatus* by multidimensional scaling with the program ALSCAL in SPSS^X (SPSS Inc. 1986). The Euclidean distance model was used to produce a non-metric multidimensional scaling plot in three dimensions (50 iterations). Kruskal's stress value and the squared correlation coefficient (SPSS Inc. 1986) indicated goodness-of-fit between the plot and the original distance matrix.

Levels of gene flow among populations were estimated using several approaches: (1) S. Wright's (1951) formula, $Nm = 1/4(1/F_{ST} - 1)$; (2) Slatkin's (1981) method based on the conditional average frequency of alleles [p(i)], i.e., the average frequency of alleles relative to the number (i) or proportion (i/d) of populations in which it occurs; and (3) Slatkin's (1985b) rare allele method, in which $\ln[p(1)] = -0.505 \ln(Nm) + (-2.44)$ and $p(1)$ is the average frequency of private or unique alleles (i.e., alleles that occur only in single populations). The value Nm represents the average number of migrants exchanged among populations per generation (product of effective population size and immigration rate). Estimates of gene flow using Slatkin's (1985b) method were obtained for single populations, for combined populations with one sample excluded, for certain subsets of populations, and for all samples combined (see N.K. Johnson and Marten 1988:184 for further discussion of these alternative approaches). Slatkin's rare allele formula is most accurate for sample sizes of 25 or greater (Slatkin 1985b). Because my sample sizes were less than 25, I applied Slatkin's (1985b:59) correction factor $(N_i/25)^{-1}$ to estimates of Nm, where N_i is the average sample size; thus, $Nm_c = Nm(N_i/25)^{-1}$. Sample sizes averaged 16.41 and the correction factor was 0.656 for the 32 populations of *P. inornatus* analyzed electrophoretically. This correction factor was adjusted for differences in sample sizes when combined populations were analyzed.

Analysis of mtDNA sequence data. DNA sequences were entered into a Macintosh computer using MacDNASYS Pro, version 1.0 (Hitachi Software Engineering America 1991), then aligned and translated according to the mammalian genetic code. A published sequence of *P. inornatus* (Edwards et al. 1991) was used initially for alignment. However, this sequence was dropped from the analysis because of two suspect substitutions at codons 41 and 42, including a first-position transversion (codon 42) that was not shared by any other *Parus* sequences examined. Apple file exchange converted the sequence data from a MacIntosh to a DOS format. Unique haplotypes were identified and analyzed using PAUP, version 3.1.1 (Swofford 1993), and MEGA, version 1.01 (Kumar et al. 1993). MEGA was used to compute basic sequence statistics (percent nucleotide composition by codon position, transitions versus transversions, silent versus replacement substitutions, percent sequence difference) and Tamura-Nei (1993) pairwise distance estimates. I preferred the Tamura-Nei algorithm over the more widely cited

Kimura (1980) 2-parameter distance method because it does not assume equal nucleotide frequencies (0.25) throughout the evolutionary process (see Kumar et al. 1993:15-30 for a discussion of distance estimates for sequence data).

All sites were included without weighting in the phylogenetic analyses. Prior to tree construction, a frequency distribution of 1000 randomly sampled trees was obtained using PAUP to test statistically the strength of "phylogenetic signal" versus random noise in the data set (for discussion of these effects, see Hillis 1991; Huelsenbeck 1991; Hillis and Huelsenbeck 1992). Relationships among the unique haplotypes of *P. inornatus*, rooted at one (*P. gambeli*) or both (*P. bicolor, P. gambeli*) outgroups, were analyzed using maximum parsimony and neighbor-joining techniques with 1000 bootstrap replications (PAUP and MEGA programs, respectively). An exhaustive parsimony analysis was performed to search for the shortest tree(s). Bootstrapping of these data was performed using the branch-and-bound method, resulting in a 50% majority rule consensus tree. The neighbor-joining tree was constructed based on Tamura-Nei (1993) distances.

Analysis of Vocal Variation

Audiospectrograms of tape-recorded songs were prepared using a Kay Elemetrics 6061B 85-16000Hz Spectrum Analyzer with the "wide-band" setting (80-8000 Hz frequency range). All songs recorded from each individual were listened to by ear and noted in a tape log, and visual displays were produced for each change in song type given by a singing male. Spectrograms were then examined and grouped into unique song types based on clearly visual differences in structural, temporal, and frequency components of notes, syllables, and entire songs. The spectrograms presented in this study were chosen to represent the range of song types recorded.

Mantel Tests

Mantel's (1967) non-parametric test compares two distance matrices and examines the null hypothesis that the matrices are randomly associated with each other; rejection of this hypothesis indicates that the pairwise distances in one matrix are significantly correlated with those in the second matrix (see discussion in Douglas and Endler 1982; Schnell et al. 1985). The Students *t*-distribution is used to test the statistical significance of the Mantel Z-statistic computed for each comparison. A correlation coefficient also is computed as a measure of the degree of concordance between matrices.

Although Mantel's procedure was developed originally to examine spatial and temporal clustering of diseases (Mantel 1967), Sokal (1979) recommended its use in studies of geographic variation and numerical taxonomy. Since then, numerous biologists have employed this approach to compare two or more data sets in evolutionary ecology (Douglas and Endler 1982), animal behavior (Schnell et al. 1985), and systematics (e.g., Zink 1986; N.K. Johnson and Marten 1992). I used Mantel's test (NTSYS-PC, version 1.50; Rohlf 1988) to assess congruence in the patterns of variation exhibited by size, color, allozymes, and mtDNA among populations of *Parus inornatus.* Seven comparisons were made: male size vs. color; female size vs. color; male size vs. allozymes; female size vs. allozymes; color vs. allozymes; and allozymes (two matrices) vs. mtDNA. The same distance measures were used as in other analyses, although matrices were reduced such that populations without both types of data were excluded from the analysis. For example, because only 32 of the 60 populations analyzed morphometrically also were studied allozymically, comparisons of size vs. allozymes were based on a 32 x 32 matrix.

Because the procedure-wise error rate (Type I error) increases when more than two matrices are compared, the probability level and associated *t*-value should be adjusted accordingly; the adjusted *P* is $0.05/n$, where *n* is the number of pairwise matrices being compared (Douglas and Endler 1982). In this study, the corrected probability level for a Type I error was $0.05/7 = 0.0071$ ($t = 2.694$; two-tailed test). Two matrices also show significant association if the probability is close to 1 that a random Mantel statistic *Z* is less than the observed *Z*.

Analysis of Environmental Correlations

Canonical correlation analysis was used to assess whether morphometric, colorimetric, or allozymic traits vary independently of environment. This method of analysis calculates pairs of canonical variates for dependent and independent variables such that the correlation between the first pair of variates is maximized (Manly 1986) . Titmice are especially suited for such an analysis because of their non-migratory behavior. Climatic data were obtained from weather stations located within 56 of the 60 analytical sample areas; such data were not available for Baja California (sample areas 28-30) and the Guadalupe Mountains (60). Twenty-six environmental variables were used for analysis: elevation, latitude, longitude, average maximum April temperature, average minimum April temperature, April mean temperature, average maximum May temperature, average minimum May temperature, May mean temperature, average maximum June temperature, average minimum June temperature, June mean temperature, average maximum December temperature, average minimum December temperature, December mean temperature, average maximum January temperature, average minimum January temperature, January mean temperature, seasonal temperature difference (= June mean temperature - December mean temperature), mean April precipitation, mean May precipitation, mean June precipitation, mean August precipitation, mean December precipitation, mean January precipitation, and mean annual precipitation. These data were obtained from monthly climatological summaries for each state (1951-1980, 1961-1990) prepared by the National Oceanic and Atmospheric Administration National Climatic Data Center (Ruffner 1985, Owenby and Ezell 1992). Canonical variates for all data sets were obtained using the MANOVA routine in SPSS[x] (SPSS Inc.1986).

SYNOPSIS OF NATURAL HISTORY

Previous studies of *Parus inornatus* have dealt with several important aspects of natural history, including breeding biology (Price 1936; Dixon 1949, 1956), molt cycle (Dixon 1962), foraging ecology and behavior (S. Davis et al. 1973; Block 1989, 1990), and singing behavior (Dixon 1969, Gaddis 1983, L.S. Johnson 1987). The most detailed analyses are those of Price (1936) and Dixon (1949, 1956), who followed marked individuals through one or more seasons in Palo Alto and Berkeley, California, respectively. Except for Block's research, all of these studies focused on single populations of titmice. Furthermore, most of the studies examined populations occupying oak-dominated habitats in California. Although these analyses have contributed significantly to information on the basic biology of titmice in western North America, additional studies are needed that compare populations with different habitat preferences (e.g., oaks versus junipers) and population densities. Studies that address geographic differences in territoriality and dispersal are particularly important because of the relevance of these issues to patterns of genetic variation and gene flow.

BREEDING BIOLOGY AND ANNUAL CYCLE

Members of the genus *Parus* are generally resident throughout their geographic distributions. Studies of several species have revealed strong sedentary behavior in adults and the tendency for permanent matings (see Hinde 1952; Matthysen 1990). These generic traits have led to important questions concerning pair formation and relationships, territoriality, nesting, juvenile dispersal, and survivorship in *P. inornatus*.

Pairing

Evidence from banding and collecting indicates that pair formation in *P. inornatus* initially occurs after the break-up of family groups in summer. Dixon (1949) banded

26

several first-year birds in August and October-November 1947 that were paired either with other immatures or with adults. Grinnell (1923) collected a pair of immature birds on 27 September 1922 near Steele Meadow, Modoc County, California. Likewise, I have collected numerous pairs from various localities that included at least one first-year bird. Although most of these birds were taken in the fall, two such pairs were collected on 27 June 1989 in northern Siskiyou County, California.

After the establishment of pair bonds, matings appear to be permanent unless one of the birds disappears. Price (1936) followed 14 banded pairs from 1928 to 1933 and showed that 11 retained mates for at least two years. He observed only one case of "divorce," in which an individual remated while its former mate was still living. Although both sexes appear to have equal status during the fall and winter months, males assert dominance and leadership during the breeding season (Dixon 1949).

Territoriality

Immature titmice establish territories in the fall, which they maintain through the winter and into the breeding season if a suitable nest cavity is available (Dixon 1949). Price (1936) and Dixon (1949) found that birds typically kept the same territory between successive years, and that individuals showed a strong tendency to reuse nest or roost sites. Males and females were equally philopatric, remaining established on their territory even after disappearance of the mate (Price 1936); these individuals usually remated by the next breeding season, sometimes pairing with wandering first-year birds (Dixon 1949). Although such sedentary behavior has important implications for the analysis of genetic population structure, the data are limited to two populations inhabiting oak woodlands in the central Coast Range of California. My field experience revealed that titmice inhabiting juniper or pinyon-juniper woodland exhibit a much greater tendency for movement than oak-associated birds. In particular, titmice in the interior often moved significantly farther distances when chased (up to 1/2 mi) than those from coastal populations, suggesting a weaker territorial association. A quantitative study of the vagility of individuals or pairs from different populations and habitats would shed further light on geographic variation in the territorial behavior of *P. inornatus*.

Intraspecific defense of territories occurs year-round and may involve either or both sexes (Dixon 1949). In late summer, defense is primarily oriented toward the exclusion of immature birds searching for space (Dixon 1956). Singing is infrequent at that time, but becomes more pronounced during the spring when the intensity of aggression heightens (Dixon 1949). The varied calls and scolding notes uttered by males and females also serve in routine territorial defense, whereas wing vibrations and chases are used during the most heated encounters (Dixon 1949).

Ecological overlap between *P. inornatus* and other sympatric parids has prompted several workers to investigate questions pertaining to interspecific territoriality. Dixon (1954) observed that *P. inornatus* and *P. rufescens* (Chestnut-backed Chickadees) occupied mutually exclusive territories where their preferred habitats intermingle in the east San Francisco Bay region of central California. Plain Titmice were dominant in all interspecific interactions, having a distinct advantage because of their larger size and year-round territorial occupancy. In Arizona and New Mexico, ecological and behavioral segregation between *P. inornatus* and *P. wollweberi* (Bridled Titmice) in areas of sympatry was taken to infer interspecific territoriality (Dixon 1950, Marshall 1957). This notion was refuted by Gaddis (1987), who found no evidence for interspecific exclusion of territories in the Chiricahua Mountains of southeastern Arizona. Breeding territories often overlapped without antagonistic interactions, and individuals of *P. inornatus* frequently joined the family flocks of *P. wollweberi* after their own family groups disbanded in late June. The two species associated in mixed flocks throughout the year except during the breeding period, presumably because of the advantage provided by increased vigilance against predators (Gaddis 1987). Marshall (1957) failed to find these two species associating during the winter or at other times in the Peloncillo Mountains of New Mexico.

Territorial boundaries were remarkably stable during a 6-year study (1947-1952) in Strawberry Canyon, Berkeley, California (Dixon 1956). Of the changes that occurred, most involved the addition of new territories in previously unoccupied habitat, resulting in an increased number of pairs nesting in the canyon. Territorial behavior probably imposes an upper limit on population density (Dixon 1956), particularly in areas such as Strawberry Canyon where the habitat is linear and the availability of new sites is limited.

The average size of 12 territories mapped in Strawberry Canyon from 1947 to 1948 was 6.3 acres (Dixon 1949); territory size ranged from 3.3 to 12.5 acres, depending on the quality of habitat. In nearby Tilden Park, 11 territories in live-oak woodland averaged 5.7 acres (Dixon 1954). Nest sites averaged one per territory for the population inhabiting Strawberry Canyon. Although first-year birds often settled in marginal habitat and seemed to choose territories without regard to nest sites, the availability of cavities dictated the distribution and density of breeding pairs. Fall-winter territories were abandoned in the spring if a suitable cavity was lacking.

Nesting Biology

I examined information on nesting, nest-site characteristics, and clutch size of *Parus inornatus* from 243 records at the Western Foundation of Vertebrate Zoology (WFVZ) and 32 records at the Museum of Vertebrate Zoology (MVZ). These data indicate that titmice will use a variety of sites for nesting, including natural cavities, woodpecker-excavated holes, and artificial boxes (Table 5). On the Pacific slope, numerous plant species in addition to tree oaks may be utilized. Furthermore, in those cases where the type of nest cavity was indicated (excluding boxes), natural holes prevailed over woodpecker excavations. The height of these nests ranged from 1.5 ft (in a hollow stump) to 40 ft, averaging 12.5 ft. Grinnell and Storer (1924) similarly found that most

TABLE 5. Summary of nest site characteristics in the *Parus inornatus* complex[a].

Type of Plant	Type of Cavity			
	Natural	Woodpecker[b]	Artificial[c]	Not Indicated
Oak	58	17	67	16
Sycamore	4	4	0	3
Cottonwood	4	1	0	0
Willow	2	6	0	0
Maple	0	0	0	1
California Bay	1	0	0	0
Elderberry	2	4	0	4
Elm	0	1	0	0
Pine (stump)	0	2	0	0
Walnut	0	1	0	0
Scrub oak	0	0	0	1
Chamise	0	0	1	0
Pinyon	0	0	11	0
Juniper	0	0	1	0
Joshua tree	0	1	1	0

[a] Based on data from 243 egg sets housed at the Western Foundation of Vertebrate Zoology. Most of the records (230) are from the Pacific slope, including 43 from northern Baja California.

[b] Included the Acorn Woodpecker (*Melanerpes formicivorus*), Ladder-backed Woodpecker (*Picoides scalaris*), Nuttall's Woodpecker (*P. nuttallii*), Downy Woodpecker (*P. pubescens turati*), and Northern Flicker (*Colaptes auratus*). Most records did not specify the type of woodpecker excavation.

[c] Mostly nest boxes, but includes two nests placed in wooden pipes attached to trees. One of these pairs originally built its nest in the eaves of a porch, but renested in the pipe after its nest was disturbed by removal of the eggs.

nests of titmice in the Yosemite region, California, were placed in naturally rotted-out cavities rather than in old woodpecker holes. Data from the interior were generally limited to birds using artificial boxes placed in pinyon. Natural cavities are scarce in pinyon, and both the density and diversity of woodpeckers is reduced compared to the oak-woodland belt of the Pacific slope. Thus, the provision of nest boxes may play an important role in enhancing local population densities of titmice in pinyon woodland. Many of the natural sites in juniper or pinyon-juniper habitat are in crevices provided by the partially decayed and twisted trunks of old juniper trees (Grinnell and Miller

1944:307). In areas where Joshua trees (*Yucca brevifolia*) co-occur with juniper or pinyon-juniper, holes excavated by Ladder-backed Woodpeckers also are used by titmice.

Females are largely responsible for nest construction (Dixon 1949). The male remains close to his mate during this period, feeding her inside the cavity and accompanying her while she gathers material. Of 193 records at the WFVZ with data on nest composition, the majority (114; 59%) were constructed with grass as a base; other materials included hair (68; 35%), moss (52; 27%), feathers (49; 25%), shredded bark (27; 14%), cotton and/or wool (23; 12%), straw (13; 7%), plant down or blossoms (9; 5%), twigs (6; 3%), twine or string (5; 3%), "plant fibers" (3; 2%), rootlets (2; 1%), snakeskin (2; 1%), wood chips (1; 0.5%), and leaves (1; 0.5%). Nest linings consisted primarily of soft material such as hair (e.g., rabbit, cow, horse, squirrel; 98 nests, 51%) and/or feathers (84; 44%).

The examination of brood patches (Price 1936; personal experience) shows that females are the sole incubators of eggs. Clutch size is moderately large, and both parents assist in feeding young. The average number of eggs in 236 sets from the WFVZ and 31 sets from the MVZ was 6.5, varying from 1 ("addled") to 9; although these data are from diverse localities, no geographic differences were detected. Price (1936) reported similar results for 62 nests at Stanford University, Palo Alto, California (average number of eggs was 6.75, with a range of 3 to 9). In keeping with the cavity-nesting habits of this species, eggs are pure white with little or no markings (see Bent 1946:415).

According to the dates of eggs deposited in the WFVZ and MVZ, the timing of incubation can vary from early March to late May or early June. The distribution of dates was as follows: 1-15 March, 1 egg set; 16-31 March, 41; 1-15 April, 83; 16-30 April, 76; 1-15 May, 46; 16-31 May, 22; 1-15 June, 1. Records from the interior and from northern Baja were generally later (late April to May) than those from the Pacific slope of California. The incubation period is approximately 14 days (Bent 1946:415). Usually one brood is reared per season.

Juvenile Dispersal

Juvenile titmice apparently disperse substantial distances from their natal territory (Price 1936; Dixon 1949), although dispersal seems to occur gradually after fledging (Dixon 1956). Of 145 immatures banded by Price (1936) at Stanford University, only two birds were ever recaptured; these were found nesting in boxes 700 and 1200 yards from the parental territory, respectively. Dixon (1949) recaptured 7 of 18 banded juveniles in Strawberry Canyon during 1947-1948, four of which survived to occupy territories at an average distance of 375 yards from their parents. Between 1947 and 1952, Dixon (1956) found that only 4 of 35 banded juveniles entered the breeding population; two others were known to settle outside of the area at distances of 550 and 800 yards from their respective birthplace. While these workers attributed the low recapture rates to dispersal, the confounding effect of high juvenile mortality also must be considered (Greenwood and Harvey 1982).

As mentioned previously, dispersing juveniles may mate with widowed individuals that are already established on territories. Otherwise, first-year birds and their mates either settle in existing vacancies or establish territories in marginal habitat. Such territories are typically abandoned in the spring if they are found unsuitable for breeding (Dixon 1949). Although immature titmice may join other birds in conspecific or mixed-species flocks during the fall and winter period (Gaddis 1987), most observations at this time are of paired individuals (personal experience).

The tendency for juvenile titmice to disperse away from their natal areas, at least in coastal California, is probably a function of high post-breeding population density and the rigid adherence of adults to their own territories through the year (Dixon 1956; see also Greenwood and Harvey 1982). Juvenile dispersal may result in the movement of surplus individuals from areas of high density ("source" populations) to areas where population density is reduced ("sinks"; see Pulliam 1988), increasing the potential for gene flow among populations. Such behavior also may act to reduce inbreeding within populations (Greenwood 1987). Once juveniles become established on favorable territories, however, the strong site-fidelity and sedentary behavior of individuals would tend to inhibit further gene flow. Additional studies of juvenile dispersal and of its consequence on gene flow are needed in regions such as the Great Basin, where population densities are generally lower and territorial behavior may be less extreme.

Survivorship

Dixon (1956:178) estimated that 76% of breeding adults survived to renest the following season in Berkeley and Palo Alto, California. Based on a value of 24% annual mortality, he determined that adults have a further average life expectancy of approximately 3.5 years. Price (1936:25) recorded 8 birds that were at least 4 years old, 3 individuals that were 5 years old, and one individual that was at least 7 years of age. Although the expected longevity of *P. inornatus* is higher than for most other small passerines (see Lack 1954:91-93), it is similar to values computed for Wrentits (*Chamaea fasciata*; Erickson 1938:309; Farner 1949:73), which also are sedentary and permanently territorial. These traits apparently confer several advantages in terms of survivorship (see Dixon 1956:178-179). The low recapture rates of juvenile titmice (Price 1936; Dixon 1949, 1956) may be due to the combined effects of high mortality and dispersal, although figures on juvenile mortality are lacking.

Molt Cycle

Adult titmice appear to have a long molt cycle that begins soon after breeding and continues into September or October. I recorded data on the timing and extent of molt (particularly of remiges and rectrices) in specimens collected during the course of this study. The earliest sign of wing molt was detected 17 May in Riverside County,

California. All specimens collected in late May and June from both coastal and interior populations were molting. In Monterey County, California, Dixon (1962) found that wing molt began in early June when fledglings were about 2 weeks out of the nest. Wing molt in that population continued into mid-September, and I have recorded it as late as 24 October in New Mexico. Titmice collected in September and early-mid October from various localities in California and the Great Basin were typically in molt, although stage of molt varied among individuals. The duration of annual post-breeding molt in titmice seems to exceed 3 months, which is similar to that of other sedentary birds such as *Carpodacus mexicanus* (Michener and Michener 1940) and *Aphelocoma coerulescens* (Pitelka 1945). However, at least for populations in central California, the molt schedule of *P. inornatus* appears to have been shifted forward so that molt terminates prior to the end of the dry season, after which arthropod food supplies become less dependable (Dixon 1962).

Dixon (1962) recognized 8 stages of wing molt in adults of *P. inornatus*, the sequence of which is the same as that reported for *P. bicolor* (Dixon 1955:129-130). Because I collected specimens at varying times in the molt cycle, I was unable to corroborate this sequence for any given population. Nonetheless, the observed patterns of replacement of primary and secondary feathers generally seemed to fall into the stages characterized by Dixon.

Dixon's (1962) analysis of molt in *P. inornatus* also revealed an incomplete post-juvenal molt, in which replacement involves the entire body plumage, all rectrices, and only some remiges (3-5 proximal secondaries and "*rarely* [italics mine] the outermost primaries and their coverts"). On the basis of this work, Pyle et al. (1987) likewise reported an incomplete post-juvenal molt. In contrast, I collected numerous immature birds with large skull windows during September and early October that seemed to be replacing all of their flight feathers (as well as tail and body feathers). Additional work is needed to clarify the extent of post-juvenal molt in this species.

FORAGING ECOLOGY

Titmice in the *Parus inornatus* complex have a varied diet, consuming large amounts of plant material in addition to insect food (Bent 1946:416). The proportion of vegetable matter in their diet increases during the fall and winter months (see A.C. Martin et al. 1951:140). At that time, individuals often use their relatively stout bill to pound acorns and other food against branches or to probe limbs for insects. Although an elevated foraging niche appears to be favored (S. Davis et al. 1973), titmice also commonly forage on the ground at the base of trees or shrubs (personal experience). Foraging ecology appears to vary seasonally and geographically (Block 1990), a finding that has important implications for interpreting patterns of morphologic variation.

GEOGRAPHIC VARIATION
IN HABITAT PREFERENCE
AND POPULATION DENSITY

Because habitat type can profoundly influence phenotypic, genetic, and vocal traits, an understanding of geographic differences in habitat preference is essential. Unfortunately, such information is often highly generalized in the literature. For example, the American Ornithologists' Union (1983:516) broadly describes *Parus inornatus* as a resident of "pinyon-juniper and oak woodland." Fieldwork during 1988-1994, however, revealed substantial local and macrogeographic variability in the composition, distribution, and relative fragmentation of woodlands within the range of the species. Associated differences in the abundance and occurrence of titmice also became evident. Thus, personal inspection of a wide representation of sites was necessary to gather firsthand data on habitat preference and population density in different geographic regions. On-site counts were obtained for individuals inhabiting continuous or fragmented woodlands dominated by oaks, junipers, and/or pinyon. The importance of other plant species also was assessed. These data, which are based on extensive fieldwork throughout the western United States, greatly enhance the scattered information currently in the literature.

Samples for analysis (Figs. 3-4, Table 2) were chosen partly to reflect the range of habitat relationships and abundances exhibited by titmice within and between each region. Numbers of titmice were typically estimated during 4-5 hours daily of observation and/or collecting. In addition to providing crucial information relevant to the genetic structure of titmouse populations, this information provides a geographic baseline of density and distribution that is invaluable for conservation biologists attempting to follow long-term trends. Furthermore, the fact that titmice are non-migratory enhances the utility of such data, because changes in numbers and distribution can clearly be ascribed to varying local or regional conditions rather than to changes at remote sites (e.g., wintering grounds). A summary of geographic variation in habitat preference and population density in titmice is presented in Table 6.

TABLE 6. Summary of geographic variation in habitat preference and population density of titmice in the western United States.

Sample Area(s)	Geographic Region	Elevation (ft)[a]	Primary Habitat[b]	Habitat Distribution[c]	Population Density[d]
1	Rogue River Valley, Southwestern Oregon	1750 (1500-2000)	Garry oak woodland/ *Ceanothus cuneatus*	continuous	high
2	Shasta Valley, Northern California	3100 (2500-3800)	Garry oak woodland mixed with juniper, *Ceanothus cuneatus*	continuous	high
4-23, 25-27	Sierra Nevada, Coast, Transverse, Peninsular ranges[e]	2600 (40-6700)[f]	oak woodland, arid oak-pine woodland	generally continuous	generally high
24	Little San Bernardino Mountains	4500 (3600-5000)	pinyon-juniper or juniper woodland with scrub oaks, Joshua trees	discontinuous	low
3, 31-33	Northwestern Great Basin, including Modoc Plateau	5200 (4250-5700)	juniper woodland	discontinuous	low to moderate
34-36	Eastern California, Western Nevada	6650 (4500-7500)	pinyon-juniper/juniper woodland	discontinuous	low to moderate
37	Eastern Mojave Desert	6000 (4400-7000)	pinyon-juniper mixed with Joshua trees	discontinuous	low
38-60	Central Great Basin to Rocky Mts., Southwestern U.S.	6100 (2250-8000)	pinyon-juniper woodland	discontinuous	low to moderate

[a] Mean elevation (range in parentheses) as determined from specimen records with elevational data (Appendix A). Elevations were averaged to the nearest 50 ft.
[b] More open woodlands are preferred over dense stands. Use of pinyon is subordinant to juniper in mixed woodland. Trees with nest cavities or crevices are essential.
[c] Continuous habitats occur widely along a particular elevational zone; discontinuous habitats are patchy and separated by areas of unsuitable vegetation (e.g., desert).
[d] Based on field observations; specific numbers are given in text. Titmouse densities in juniper or pinyon-juniper woodland are, on average, approximately 25-50% lower than in oak-dominated woodlands. Three areas (Joshua Tree [24], Eastern Mojave [37], Reserve [56]) showed an unexpected decline in density from previous records.
[e] Locally, certain populations occupy discontinuous habitat of pinyons mixed with oaks, gray pine, and/or Joshua trees. Densities are lower than in oak woodland.
[f] Elevational distributions are generally higher in the southern part of the region: for sample areas 4-15, the mean elevation is 1550 ft (range = 40-4500 ft); for sample areas 16-27, the mean elevation is 4000 ft (range = 400-6700 ft).

ROGUE RIVER VALLEY, SOUTHWESTERN OREGON
(sample area 1)

Titmice are common in the warm, dry woodlands of Garry oak (*Quercus garryana*) mixed with buckbrush (*Ceanothus cuneatus*) (Fig. 5) that surround much of the Rogue River Valley in Jackson County, Oregon. Associated plants include California black oak (*Q. kelloggii*), ponderosa pine (*Pinus ponderosa*), grayleaf manzanita (*Arctostaphylos viscida*), and/or poison oak (*Rhus diversiloba*). This oak-chaparral community dominates the slopes from the floor of the valley to approximately 2500 ft elevation (Browning 1975). I observed 6-8 singing males, pairs, or family groups daily in this region during June and September 1991.

FIGURE 5. Garry oak (*Quercus garryana*) and buckbrush (*Ceanothus cuneatus*) habitat occupied by titmice in the Rogue River Valley, southwestern Oregon (sample area 1). Ponderosa pines (*Pinus ponderosa*) are scattered throughout (shown in background). Note the grassy understory and the openness of arboreal vegetation.

SHASTA VALLEY, NORTHERN CALIFORNIA
(sample area 2)

Shasta Valley, Siskiyou County, California, is separated from the Rogue River Valley by coniferous forest that occurs at higher elevations in the Siskiyou Mountains. Grinnell and Miller (1944:307) described the habitat of *Parus inornatus sequestratus* in this area as consisting of a "mixed association of juniper, Garry oak and *Ceanothus cuneatus* and Garry oak woodland." These authors reported the occurrence of localized populations of *sequestratus* from Shasta Valley and from the South Fork of the Trinity River in Trinity County (and "probably.... in intervening valleys"). In 1989-1990, I found titmice to be abundant in such habitat east and northeast of Shasta Valley (15-25 individuals seen per morning).

Southeast of the valley, titmice occupy a strikingly different woodland dominated by large western junipers (*Juniperus occidentalis*) with an understory of big sagebrush (*Artemesia tridentata*), manzanita, and antelope bitterbrush (*Purshia tridentata*). Scattered yellow pines also occur, but oaks are conspicuously absent. Such habitat typically characterizes populations farther east in the western Great Basin. I collected in this area during 1990-1991 and encountered 4-5 pairs or family groups each morning. Density was notably lower here than in the oak-juniper-chaparral association occupied elsewhere near Shasta Valley.

SIERRA NEVADA, COAST, TRANSVERSE, AND PENINSULAR RANGES
(sample areas 4-23, 25-27)

Titmice are common residents of oak or oak-pine woodlands (Fig. 6) that occur widely throughout the western foothills of the Sierra Nevada, in the Coast Ranges, and on the Pacific slopes of the Transverse and Peninsular ranges in southern California. Although present in the Sacramento Valley, including the isolated Marysville Buttes, they are absent from the more arid San Joaquin Valley to the south. Eighteen species of tree or shrub oaks are currently found in California (Pavlik et al. 1991). Pure or mixed stands of tree oaks may consist of valley oak (*Quercus lobata*), California black oak, blue oak (*Q. douglasii*), Garry oak, Engelmann oak (*Q. engelmannii*), coast live oak (*Q. agrifolia*), interior live oak (*Q. wislizenii*), canyon live oak (*Q. chrysolepis*), and/or interspecific hybrids (Griffin and Critchfield 1972; Pavlik et al. 1991). In some areas, gray pine (*Pinus sabiniana*), ponderosa pine, and/or Coulter pine (*P. coulteri*) co-occur to form a mixed oak-pine woodland. Other tree species may include California buckeye (*Aesculus californica*), madrone (*Arbutus menziesii*), and tanoak (*Lithocarpus densiflorus*). Understory vegetation is characterized by herbaceous plants (especially annual grasses) and shrubs such as toyon (*Heteromeles arbutifolia*), buckbrush, coffeeberry (*Rhamnus californica*), poison oak, and manzanita.

FIGURE 6. Typical oak-pine habitat occupied by titmice along the Pacific slope in California (photograph taken in Yolo County [sample area 5]). Vegetation here consists predominantly of blue oak (*Quercus douglasii*) mixed with gray pine (*Pinus sabiniana*) and an herbaceous understory.

Geographic variation in plant-species composition within the oak-woodland zone results in differential patterns of habitat use by *Parus inornatus* (Block 1990). In general, however, titmice prefer fairly open oak woodlands or savannas with warm microclimates (personal experience). Scrub oak or other brush may be utilized if woodland occurs nearby. On San Benito Mountain in the southern Diablo Range, San Benito and Fresno counties, pine forests (*Pinus sabiniana*, *P. coulteri*, *P. jeffreyi*) lacking arboreal oaks also are utilized (N.K. Johnson and Cicero 1985:8). Titmice were abundant in all areas of oak or oak-pine woodland where I collected between 1989 and 1991 (sample areas 4-6, 10, 12, 15, 18, 27). At least 8-10 pairs or groups were typically observed each morning.

Titmice occupy a different habitat in the southeastern Sierra Nevada and on the desert side of the Transverse and Peninsular ranges, where coastal and interior biotas meet across a sharp environmental gradient. Vegetation in this ecotone contains elements of both floras that include singleleaf pinyon (*Pinus monophylla*), western juniper, Joshua trees, gray or Coulter pine, canyon live oak, and/or various chaparral or desert-shrub species. During fieldwork in 1989-1990 near Walker Pass, Kern County, California (sample area 16), I found titmice uncommonly in pinyon mixed with canyon live oak and gray pine above 5600 ft (3-4 pairs seen daily). Fewer individuals (1-2 per day) were observed below that zone in pinyons mixed with Joshua trees, and no titmice were seen in areas of pure pinyon. Southwest of Walker Pass, the primary habitat of *Parus inornatus* consists of oak-pine woodland typical of coastal populations. Singleleaf pinyon grows uncommonly on the ridges there (R.M. Gilmore, unpubl. field notes in MVZ archives), extending westward through the Tehachapi Mountains to the Coast Range, where it occurs in large dense stands in northern Ventura County (Griffin and Critchfield 1972). No records of titmice exist from pinyon woodland in that area.

The San Bernardino and San Jacinto mountains in southern California form another strong break between the Pacific Coast region and the Mojave Desert. Although I did not collect any specimens from these ranges, they warrant discussion because of the mixture of coastal and interior habitats available to titmice. The dominant life zone in the San Bernardino Mountains is Upper Sonoran, consisting of chaparral, oak or oak-pine woodland, and pinyon-juniper (see Grinnell 1908). While most records of titmice are from the west-southwest slope, the species has been observed and collected on the desert side in canyons grown to *Quercus chrysolepis*, pinyon-juniper, and tree yuccas (Grinnell 1908). Egg sets (WFVZ) also have been taken from pairs nesting in artificial boxes or in woodpecker cavities in Joshua trees and junipers on the northeast slope of the range near Hesperia. In June 1989, I found no titmice in pinyon mixed with scattered junipers and Joshua trees on the east slope of the San Bernardino Mountains. Although titmice also occur on both the desert and Pacific slopes of the San Jacinto and Santa Rosa mountains, they were most abundant on the desert side of the San Jacinto Range during a faunal survey in 1908 (Grinnell and Swarth 1913:310).

LITTLE SAN BERNARDINO MOUNTAINS
(sample area 24)

The Little San Bernardino Mountains run through the heart of Joshua Tree National Monument in Riverside and San Bernardino counties, California. They are separated from the main San Bernardino Mountains by Morongo Valley, with a narrow connection across Morongo Pass (A.H. Miller 1946). The Monument lies at the junction of the Mojave and Colorado deserts, and thus the environment is decidedly arid. Nonetheless, scattered tracts of pinyon and California juniper (*Juniperus californica*) occur along the crest and upper north-facing slopes of the Little San Bernardino Mountains. The woodland is mixed with desert scrub and/or chaparral comprised of scrub oak (*Quercus turbinella*), manzanita, desert bitterbrush, and mountain mahogany (*Cercocarpus* sp.). Joshua trees are interpersed through the woodland in some areas.

FIGURE 7. California junipers (*Juniperus californica*) and Joshua trees (*Yucca brevifolia*) mixed with desert scrub at Juniper Flat, Little San Bernardino Mountains, San Bernardino County, California (sample area 24). A pair of titmice was found nesting here in an old Ladder-backed Woodpecker (*Picoides scalaris*) cavity excavated in a Joshua tree.

A.H. Miller and Stebbins (1964:148) described *Parus inornatus* as a "common resident of [the] upper levels of [the] western section" of the Monument. At Pinyon Wells, 4000 ft (the type locality of *P. i. mohavensis*), A.H. Miller (unpubl. field notes in MVZ archives) reported titmice to be "abundant in the pinyons from 4200 to 5000 ft on the crest of the ridge.... probably a total of 50 [single or paired individuals] were detected in the course of the morning [13 October 1945]." Titmice also were recorded from several other localities in the Monument, including Black Rock Canyon, where numerous specimens were taken. Although most individuals were observed above 4000 ft in pinyon, juniper, and scrub oak, small numbers were found locally to 3700 ft, as at Quail Spring (A.H. Miller and Stebbins 1964:149).

I made 4 trips to Joshua Tree National Monument during 1989 and 1991 (sample area 24). Although most of my time was spent in Black Rock Canyon, 4000-5000 ft, I visited several other localities where titmice had been reported previously. Given the earlier records of abundance, I was surprised at the low density of titmice. On average, only 1-3 pairs were observed daily, and many areas of apparently suitable woodland were devoid of titmice. The presence of scrub oaks did not seem to influence the distribution or density of titmice. At Juniper Flat, 4800 ft, titmice frequented small to medium-sized junipers and Joshua trees (Fig. 7) extensively but were scarce or absent from adjacent slopes dominated by pinyon, juniper, and scrub oak. The use of cavities in Joshua trees may dictate the distribution of titmice here, at least during the nesting season.

On 19 May 1989, I spent several hours searching for titmice near Pinyon Wells. Woodland was localized in draws on north-facing slopes, and much of the area seemed unsuitable; no individuals were seen during the course of the morning. I cannot explain the great disparity in density of titmice between my visit and that of A.H. Miller in 1945.

BAJA CALIFORNIA, MEXICO
(sample areas 28-30)

Records of *Parus inornatus* from northern Baja California extend from the United States border to the southern Sierra San Pedro Mártir. According to Wilbur (1987:124), the species is common there in riparian and chaparral habitats which are part of the "Californian" botanic region. The flora of this association includes *Quercus engelmannii, Q. agrifolia, Q. dumosa, Adenostoma fasciculatum, A. sparsifolium, Ceanothus,* and *Arctostaphylos* (Wilbur 1987:10). Titmice also inhabit single-leaf pinyon woodland on the upper slopes of the Sierra Juarez and Sierra San Pedro Mártir. The distantly isolated form in southern Baja California is a common resident of the higher mountains of the Cape District (Grinnell 1928:223). Upper Sonoran vegetation in the Sierra de la Laguna consists of several species of pines (*Pinus cembroides, P. edulis*) and oaks (*Q. brandegei, Q. peninsularis, Q. devia*) associated with the Arid Tropical botanic region (Wilbur 1987:11).

NORTHWESTERN GREAT BASIN
(sample areas 3 and 31-33)

The distribution of *Parus inornatus* in the northwestern Great Basin extends from northeastern California to scattered localities in northwestern Nevada (east side Granite Mountain, Mud Lake) and southeastern Oregon (Warner Valley, Hart Mountain National Antelope Refuge [T.H. Rogers 1985:940], northwest slope Steens Mountain). A major physiographic feature of this region is the Modoc Plateau, an extensive area east of the Cascade Range that is characterized by western juniper woodland, sagebrush flats, and lava flows. Other plant associates include rabbitbrush (*Chrysothamnus*), antelope bitterbrush, and mountain mahogany. Coniferous forest occurs at higher elevations around the margins of the Plateau. Grinnell (1923:135) noted the failure of W. P. Taylor and associates to find any titmice on the Modoc Plateau during three months of fieldwork there in 1910. Subsequently, Grinnell collected a pair from Steele Meadow (near Clear Lake, Modoc County) on 27 September 1922. Using these specimens and other scattered records, Grinnell and Miller (1944:306-308) pointed to Clear Lake as the probable western limit of titmice in the Modoc region. Accordingly, they showed a gap of approximately 50 mi (83 km) between *Parus inornatus zaleptus* in northeastern California and *P. i. sequestratus* in Shasta Valley.

Although juniper woodland is most extensive south and east-southeast of Clear Lake, patches of juniper separated by sagebrush, grassland, agriculture, or wetland occur westward from Clear Lake through the Modoc Plateau to the Cascade Range. I found titmice occupying several patches within this presumed gap (sample area 3; Fig. 8) during fieldwork in 1990-1992. Specimens were collected from two sites at the western edge of the Modoc Plateau (vicinity of Dorris and Red Rock Valley, 4400 ft) and from one locality just west-northwest of Lava Beds National Monument (southwest slope of Mt. Dome, 4200-4300 ft). Densities were higher at Mt. Dome (4-7 pairs seen daily) compared to the other two populations (1-3 pairs seen daily), although encounters at all sites were widely spaced. Junipers become small and spottily distributed further east, where extensive lava flows separate the habitat near Mt. Dome from the more continuous woodlands south and east of Clear Lake. While pockets of juniper occur locally within the Monument, more substantial woodland grows at the southern edge of the lava. Coniferous forest dominated by ponderosa pine occurs to the south, and junipers mix with scattered pines at the forest-woodland ecotone. I obtained the first records of titmice from this area in September 1991, when 4 pairs were encountered near the southern entrance to the Monument. Subsequent trips during the breeding season revealed lower population densities (1-2 pairs seen daily). Specimens from this locality were excluded from the present study because they were taken after most analyses were completed.

The patchy association of junipers and pines, varying from pure to mixed assemblages, continues southeastward from Lava Beds National Monument. Gradually, the mixture gives way to more extensive stands of juniper woodland that typify northeastern California. Because this area represents the previously established western limit of titmice in the region, I made several trips here during 1990-1991 to collect a small series for analysis (sample area 32). Titmice occurred in low density, with only 1-3 individuals or pairs seen daily. Although titmice were more common east and southeast of the Modoc Plateau, pairs were still widely spaced compared to those occupying oak woodlands of the Pacific slope. In southeastern Oregon (sample area 31) and northeastern California (sample area 33), 3-6 pairs or family groups were seen daily in June 1989 at intervals of approximately 0.5 mi. The availability of large junipers and snags undoubtedly influences local densities of titmice in these areas.

FIGURE 8. Western juniper (*Juniperus occidentalis*)-sagebrush (*Artemesia tridentata*) habitat occupied by titmice near Mt. Dome (sample area 3), Siskiyou County, California, on the Modoc Plateau. View is looking southwest from Mt. Dome toward Mt. Shasta. Juniper woodland occurs patchily from this area eastward to Clear Lake, Modoc County. More continuous woodland to the west is visible in background. Extensive woodland also occurs east of Clear Lake in northeastern California.

EASTERN CALIFORNIA AND WESTERN NEVADA
(sample areas 34-36)

A major contrast between this region and northeastern California is the occurrence of single-leaf pinyon, which reaches its northern limit near the Humboldt River in Nevada. South of Reno in the western Great Basin, pinyon-juniper woodland dominates the slopes of desert mountain ranges at mid-elevations. The composition of the woodland differs locally and geographically, varying from nearly pure to mixed stands of pinyon, western juniper, and/or Utah juniper (*J. osteosperma*). Similarly, the habitat occupied by titmice ranges from pure juniper in the north and locally at other sites to a more diverse pinyon-juniper woodland. Population densities of titmice also vary, and are strongly reduced in some areas. Titmice are most numerous in woodlands dominated by juniper, which is favored over pinyon. Dense pinyon on cool, north-facing slopes is avoided.

Prior to my work, titmice had been recorded and collected from scattered localities throughout eastern California and western Nevada (west side of Pyramid Lake, Washoe County, Nevada, south to the Panamint Range, Inyo County, California). Subsequently, low densities have been found in several disjunct mountain ranges further east in western Nevada. Of these sites, the largest stand of isolated woodland occurs in the East Range, Pershing County, where 1-2 pairs were observed in pure juniper woodland on each of two visits during June and October 1991. This locality is approximately 85 mi (137 km) from the nearest previously known population in the Granite Range, Washoe County, Nevada. Another new record was obtained in June 1993 from the west slope of the Jackson Mountains, Humboldt County, Nevada (N.K. Johnson, unpubl. field notes in MVZ archives), about halfway between the Granite and East ranges. Farther south, titmice were scarce in pinyon-juniper woodland at the east base of Fairview Peak, Churchill County, Nevada (1 pair seen in September 1990), and in the Pilot Mountains, Mineral County, Nevada (1-3 widely scattered individuals seen daily in May and June 1990 and 1991). These outposts occur approximately 60 and 40 mi (97 and 65 km), respectively, from the nearest populations to the west.

The patchiness of titmice in this region is exemplified by records obtained during 1990-1991 in the White-Inyo Mountains, Inyo County, California (sample area 36). Because of the broad elevational span occupied by pinyon-juniper here (St. Andre et al. 1965), titmice and other non-boreal species occur at higher elevations than in comparable environments elsewhere (e.g., Panamint Mountains; see N.K. Johnson and Cicero 1986). Although titmice were fairly common near Westgard Pass on 25 June 1990 (4-5 pairs or family groups observed), only one family group and a single male were recorded the following day in similar habitat. No titmice were seen on 27 June 1990 in 5 hours of hunting. Titmice also occurred sparsely through the woodland during the spring and fall of 1991, with a maximum of 1-2 encounters daily.

EASTERN MOJAVE DESERT
(sample area 37)

Field parties from the Museum of Vertebrate Zoology surveyed the Providence Mountains and adjacent isolated ranges of the eastern Mojave Desert, San Bernardino County, California, periodically between 1917 and 1945. They collected a total of 68 specimens of *P. inornatus*, most of which were taken in winter when population densities are typically higher. However, titmice were recorded in small numbers year-round "at most localities where there were pinons or junipers." These workers found titmice to be "exceptionally abundant" in Cedar Canyon in the Providence Mountains, presumably because of the availability of Joshua trees for nesting (D.H. Johnson et al. 1948:304-305).

In contrast, when I visited Cedar Canyon on 11 April 1991, only 4-5 widely-spaced individuals were encountered on the south-facing slope of the canyon, 5600-5800 ft. Although Joshua trees were prevalent in the main wash and along the base of the slopes, they occurred sparsely through the woodland at higher elevations. On Clark Mountain, where D.H. Johnson et al. (1948) found titmice uncommonly on the north and southeast slopes, I observed one family group on 9 June 1989 in small pinyons and junipers mixed with desert scrub, yuccas, and cactus.

CENTRAL GREAT BASIN TO ROCKY MOUNTAINS, SOUTHWESTERN UNITED STATES
(sample areas 38-60)

Titmice occur patchily throughout this broad geographic region, ranging as far north as the Snake River drainage in southern Idaho. Recently, breeding birds also have been recorded near the Green River in southwestern Wyoming (Kingery 1981:965, 1982:1001). Occurrence is particularly spotty in the central Great Basin, where populations are restricted to disjunct stands of pinyon-juniper on the warmer slopes of isolated mountain ranges. Although the woodland becomes more continuous toward the east, intervening areas of unsuitable habitat (e.g., sagebrush desert, coniferous forest) undoubtedly act as barriers between resident populations of titmice. The lack of continuous habitat has important implications for the genetic structure of these populations.

The primary woodland species in Nevada, western Utah, southern Idaho, and southwestern Wyoming are *Pinus monophylla* and/or *Juniperus osteosperma*. *Pinus edulis* is the dominant pinyon through most of Utah, Colorado, Arizona, New Mexico, and western Oklahoma, where it mixes with *J. osteosperma* or *J. monosperma*. In southeastern Arizona and southwestern New Mexico, additional species include *P. cembroides*, *J. deppeana*, and *J. erythrocarpa*. Mountain mahogany and/or Gambel oak (*Quercus gambeli*) are common associates of pinyon-juniper, with oaks becoming increasingly diverse and abundant in the Southwest. Unlike populations along the Pacific slope, Plain Titmice do not depend on oaks in this region. However, oaks mixed with pinyon-juniper may be utilized (Marshall 1957; Gaddis 1987) as long as the habitat is fairly open and sunny (Dixon 1950).

Fieldwork in several disjunct mountain ranges in southeastern Nevada (sample area 39) revealed low densities there compared to other areas in the region (maximum of 3 pairs or family groups seen daily). The birds clearly preferred open, juniper-dominated sites, and no titmice were seen on slopes grown to denser woodland dominated by pinyon. Abundance was higher in the Ruby Mountains of northeastern Nevada (sample area 40), where at least 4 pairs or singing males were observed daily in mixed woodland near Overland Pass at the southern end of the range. According to Linsdale (1938c:88), "Oberholser and Bailey (MS) [also] found [them] common in June, 1888, at Hastings Pass, Ruby Mountains, Elko County."

The scarcity or absence of titmice in mountain ranges of central Nevada was noted by

Linsdale (1938a:38), who failed to obtain a single specimen "between 116° and 119° W longitude." Linsdale (1938b) also did not record titmice in the Toiyabe Mountains during 321 days of fieldwork there between 1930 and 1933. Likewise, subsequent workers have failed to record the species from the Toiyabe Mountains or from other high ranges in central Nevada (N. K. Johnson, pers. comm.; see also Alcorn 1988:256-257). I did not observe any titmice during several visits to central Nevada in 1990-1991. Areas of exploration included: east slope of the Reveille Range, 6200-7100 ft, Nye County; east side of the Paradise Range, 6900 ft, Nye County; southeast slope of Mt. Callaghan, 6700 ft, at the north end of the Toiyabe Mountains, Nye County; and south slope of the Roberts Mountains, 6800-7000 ft, Eureka County. The prevalence of pinyon over juniper, the small stature of available junipers, and the extreme environmental conditions, may account for their absence in the Toiyabe Mountains (Linsdale 1938b:1-14) and adjacent ranges. In contrast, titmice are resident in pinyon-juniper woodland on the northeast slope of the Monitor Range, 6400-7100 ft, Eureka County, where I obtained the first record for the species in June 1994 (1-3 singing males daily). Unlike other mountain ranges in central Nevada, juniper woodland occurs extensively here on the lower slopes. The distribution and density of juniper increase significantly to the east, as do numbers of titmice. Thus, the Monitor Range may contain the westernmost outpost of titmice in east-central Nevada. The closest known population to the west occurs approximately 100 mi (160 km) away at Fairview Peak, Churchill County, Nevada.

Population densities in Utah were similar to those recorded for eastern Nevada. In the Stansbury Mountains, Tooele County (sample area 42), the habitat occupied by titmice consists of Utah juniper mixed with mountain mahogany and various shrubs; pinyon is notably absent. Given the prevalence of juniper there, I expected titmice to be more common (only 2-4 encounters daily in June and September 1990). Titmice were seen 4-7 times daily in October 1991 during hikes of 4-6 mi in mixed pinyon-juniper on the west slope of the La Sal Mountains, La Sal County (sample area 44).

Local abundances were higher in Colorado (sample area 45) and parts of New Mexico (sample areas 48-49, 59), where 7-15 individuals or pairs were seen daily in mixed or juniper-dominated woodland. In the Jicarilla Mountains of south-central New Mexico (sample area 59), titmice were found in two strikingly different habitat formations: a low, scrubby growth; and a true arboreal woodland reminiscent of California's oak zone, with tall trees and an herbaceous understory. While Utah juniper and pinyon co-dominate both habitats, alligator juniper occurs only in the latter. Other plants associated with either type include scrub oak, ocotillo, Spanish bayonet, and various cacti. In contrast to these sites, density was much lower near Reserve in west-central New Mexico (sample area 56). This was surprising, given the huge series of titmice collected there by H. H. Kimball between 1925 and 1929. Only 2-3 pairs or family groups were observed daily during each of several visits in the spring and fall of 1990 and 1991.

Abundance in Arizona was highest on the northeast slope of the San Francisco Mountains, 6200-6400 ft, Coconino County (sample area 53; Fig. 9), where 5-10 pairs

FIGURE 9. Pinyon-juniper woodland on Deadman Flat northeast of the San Francisco Mountains, Coconino County, Arizona. Titmice were fairly common here compared to other interior localities, presumably because of the dominance of juniper and the open spacing of the woodland.

were seen daily in October 1989. In contrast, only 1-2 pairs daily were observed during April-May 1990 in the foothills of the Kaibab Plateau, 6100-6500 ft, Coconino County (sample area 52), where they showed a strong preference for sunny slopes grown to open woodland. Likewise, relatively low densities were recorded in the Chiricahua Mountains, 5400-6200 ft, Cochise County (sample area 59). In a study of ecological overlap between *P. inornatus* and *P. wollweberi* in the Chiricahua Mountains, Gaddis (1987) concluded that the occurrence of each species depends on the distribution of its preferred habitat (i.e., pinyon-juniper versus oaks). Nonetheless, he found both species breeding in areas of oaks mixed with junipers. *Parus inornatus* reaches its limit slightly to the south, where pinyon-juniper is pinched out by encinal and *P. wollweberi* becomes more numerous. I found *P. inornatus* most commonly in stands of pure pinyon-juniper without oaks during visits to this region in 1989-1990. Although I collected 3 specimens in October 1989 from oaks mixed with junipers, those were the only Plain Titmice observed in such habitat during two days of fieldwork. On the other hand, 3-4 pairs were encountered daily during the spring and fall of 1989-1990 in pinyon-juniper woodland.

PHENOTYPIC VARIATION

NON-GEOGRAPHIC VARIATION

Two major sources of intraspecific variation include variability within populations (non-geographic variation) and spatial variation among populations. A third component, temporal variability (Burns and Zink 1990; Bates and Zink 1992), also is important but will not be discussed. Non-geographic variation can have potentially confounding influence on geographic patterns (Thorpe 1976). Aspects of non-geographic variation include: (1) character correlations; (2) relative variability of different traits; (3) secondary sexual dimorphism; and (4) ontogenetic differences.

Character Correlations

The explanatory power of multivariate data sets in studies of geographic variation depends on two main issues: character redundancy, in which characters are highly correlated with one another statistically *at the level of individuals*; and character concordance, in which independent characters exhibit similar patterns of variation geographically. Characters that show serious redundancy in information should be excluded from analyses of geographic variation (e.g., see N.K. Johnson 1980:24-25). Figures 10 and 11 illustrate the correlation of size and color characters, respectively, within individuals of *Parus inornatus*. Correlation coefficients were low to moderate in all cases, indicating that each character imparts at least some unique information. Thus, all characters were retained for further analysis.

Wing and tail length were the most highly correlated size characters (Fig. 10) for both males ($r = 0.738$) and females ($r = 0.727$). Toe measurements clustered separately from other linear dimensions, with the exception of female body weight (cube root). For color (Fig. 11), purity of the breast and belly appeared to be most highly correlated ($r = 0.747$). Measures of purity and brightness clustered neatly together, although these two sets of

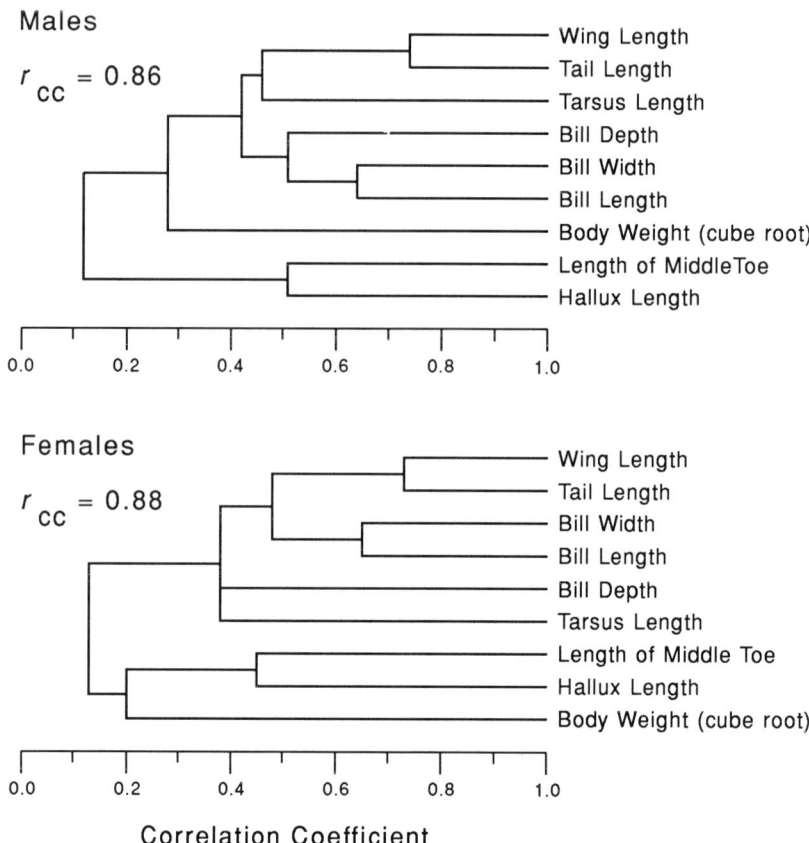

FIGURE 10. UPGMA phenograms illustrating the degree of correlation among different size characters for males (top) and females (bottom). The high cophenetic coefficients (r_{cc}) indicate strong agreement between the clustering diagrams and the original pairwise correlation matrices.

characters were uncorrelated. Dominant wavelength also clustered separately, except that dominant wavelength of the dorsum was more strongly correlated with purity than with the other wavelength characters. In all parameters of color, measurements of the dorsal and ventral plumage were only weakly correlated.

Relative Variability of Characters

Several ornithologists have investigated the question of whether different characters and taxa exhibit similar levels of morphologic variability within species. Species that

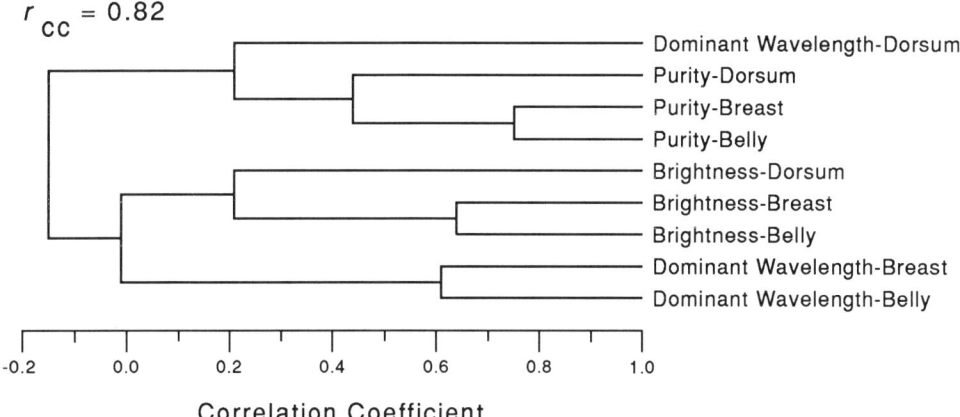

FIGURE 11. UPGMA phenogram illustrating the degree of correlation among different color variables (males and females combined). The cophenetic correlation (r_{cc}) indicates moderate agreement with the original pairwise correlation matrix.

have been examined include highly differentiated forms (e.g., *Passerella iliaca*, Zink 1986), morphologically similar sibling species (e.g., *Empidonax*, N.K. Johnson 1980), and both migratory and resident taxa. The issue of relative variability has two major implications for studies of geographic variation. First, characters that exhibit high variability within populations may contribute less to explaining the variance among populations. Such variation may be real or may be attributed to errors associated with measurement (Lougheed et al. 1991). On the other hand, character variation is the basis upon which natural selection or drift operate to produce microevolutionary changes within populations (although, as Zink [1986:82] noted, other processes may lead to similar results). Furthermore, reduced variability of characters may have fitness advantages (Simpson 1944; Bird et al. 1981).

Intra-population variation in size was generally small for *Parus inornatus*, ranging from 2.36% to 6.15% depending on the character (Table 7). The two most commonly used measures of overall size (wing length and body mass) were at opposite ends of the spectrum, with body weight being 2 to 3 times more variable than wing length. Bill dimensions were moderately variable, having slightly higher coefficients of variation than leg and toe measurements. Studies of other bird species (e.g., N.K. Johnson 1980; Zink 1986) have shown similar trends, suggesting that patterns of variation are not randomly distributed among characters. Intraspecific coefficients of variation did not differ appreciably between the sexes, a finding also reported by N.K. Johnson (1980).

TABLE 7. Relative variability of size characters in *Parus inornatus* (this study) and *Parus gambeli* (Behle 1956).

Size Character	Sex	Mean Coefficients of Variation	
		Parus inornatus[a]	*Parus gambeli*[b]
Wing Length	Males	2.36	2.90
	Females	2.42	2.98
Tail Length	Males	3.28	3.70
	Females	3.01	3.44
Tarsus Length	Males	3.01	----
	Females	3.05	----
Length of Middle Toe	Males	3.86	----
	Females	3.61	----
Hallux Length	Males	4.26	----
	Females	4.42	----
Bill Depth	Males	4.72	6.77
	Females	4.65	7.29
Bill Width	Males	4.95	6.78
	Females	4.89	7.15
Bill Length	Males	4.26	5.23
	Females	3.99	5.17
Body Weight (gms)[c]	Males	5.33	7.11
	Females	6.15	8.94

[a] Based on 60 samples distributed throughout the range of the species (except for body weight, which is based on 47 samples for males and 41 samples for females) . See Table 2 for sample sizes; average sample size for males is 26 and for females is 18.

[b] Calculated from Tables 1-7 of Behle (1956). Coefficients of variation were based on measurements pooled over the entire ranges of subspecies in western North America. Sample sizes varied from 4 to 221 depending on the sex and character.

[c] Coefficients of variation for *P. inornatus* were determined for actual weights (rather than cube roots) in order to be comparable to values for *P. gambeli*.

Coefficients of size variation were smaller for *P. inornatus* than for *P. gambeli* (Table 7), another resident parid that overlaps geographically and ecologically with *P. inornatus* through parts of its range. Wing and tail length were less variable than either body weight or bill characters within *P. gambeli*, and likewise showed smaller interspecific contrasts. Sexual differences were similar when comparing variability within the two species. The higher coefficients of variation for *P. gambeli* may be explained by the method of analysis, in which measurements were pooled over the entire distribution of each subspecies in western North America (see Behle 1956). A more direct comparison would require the study of morphologic variation in *P. gambeli* at the level of populations rather than subspecies. Coefficients of variation were slightly higher for *P. inornatus* than for *Empidonax difficilis* or *E. flavescens*, although they were similar to values reported for *E. hammondii* (N.K. Johnson 1980:33). They fall within the range of values given by Zink (1986:83) for different populations of *Passerella iliaca*.

Information on relative variability of color parameters is much more limited than it is for size, largely because of the few quantitative studies of color variation. N.K. Johnson (1980) measured color differences within and among *Empidonax* species but did not provide data on intra-population variability. His analysis of inter-population variability, however, revealed differences among traits in the extent of geographic variation: brightness and purity of the breast showed the greatest variability among populations of *Empidonax*, brightness and purity of the dorsum were moderately variable, and dominant wavelength values for both the back and breast were essentially invariant.

I calculated average coefficients of variation within populations of *Parus inornatus* for 9 color characters and compared them to values computed for 2 color morphs of *Chlorospingus pileatus* (Table 8). These results showed a clear trend in both species among the different parameters of color. Dominant wavelength was highly stable, with values comparable to those given by N.K. Johnson (1980:38) for his inter-population comparisons. Variability within populations was higher for purity than for brightness, and for each of those two measures the dorsum was less variable than either the breast (intermediate) or belly (most variable). With the exception of dominant wavelength, relative variability was substantially higher for color than for size in *P. inornatus*. Additional studies are needed to evaluate the significance of such trends.

Secondary Sexual Dimorphism

Intraspecific size dimorphism between sexes presumably reflects Darwinian sexual selection. Males are larger than females in most bird species, and this is thought to confer an advantage in terms of competition for mates (Amadon 1959). Increased size is also advantageous for territorial behavior, with larger males being more effective at defending resources (including mates) against intruders. The relationship between male size and mate competition may be weaker in *Parus inornatus* than in other passerines because of the tendency to mate permanently and because individuals secure mates during the first summer of life when they may still be growing. On the other hand, larger size may have a significant advantage in relation to territoriality because of the need to defend territories year-round.

Males of *P. inornatus* were larger than females in all characters (Fig. 12), with the percentage difference varying from 1.4% (bill length) to 6.0% (body weight). Bill

TABLE 8.   Relative variability of color characters in *Parus inornatus* (this study) and *Chlorospingus pileatus* (N.K. Johnson and Brush 1972).

| | | | Mean Coefficients of Variation | |
| | | | *Chlorospingus pileatus*[b] | |
Body Region	Color Parameter	*Parus inornatus*[a]	Yellow-Green Morph	Gray-Green Morph
Dorsum	Dominant Wavelength	0.12	0.07	0.09
	Brightness	9.86	7.14	7.85
	Purity	12.09	7.52	12.60
Breast	Dominant Wavelength	0.11	0.08	0.17
	Brightness	10.80	11.18	7.19
	Purity	16.35	14.19	31.59
Belly	Dominant Wavelength	0.11	-----	-----
	Brightness	12.66	-----	-----
	Purity	15.55	-----	-----

[a] Based on a maximum of 45 samples ($n > 1$; see Table 2); fewer samples were analyzed for variability in breast (42) and belly (39) color because of missing data due to ruffled, soiled, or open plumage. Sample sizes varied from 2 to 38 depending on the population and body region (average = 9.2).

[b] Calculated from Table 3 of N.K. Johnson and Brush (1972). Coefficients of variation are based on measurements of each of two color morphs in central Costa Rica; average sample size is 16.

characters were least dimorphic, suggesting that the two sexes do not have strong niche partitioning with regard to food resources. Unfortunately, there are no data available to support this contention. Martin and Pitochelli (1991) found that Blue Tits (*P. caeruleus*) had sexually dimorphic bills *only* where other species of *Parus* were absent, and argued that interspecific competition constrains the evolution of sexual dimorphism in bill morphology. Likewise, Gosler and Carruthers (1994) reported increased sexual dimorphism in bill shape of Coal Tits (*P. ater*) in areas where the species is socially dominant because of reduced competition. Further studies are needed to document sexual differences in the diet and/or foraging ecology of *P. inornatus* in the presence and absence of congeners.

Multivariate analysis-of-variance (MANOVA) showed a strong influence of sex on patterns of inter-population size variation in *P. inornatus* for all characters ($P < 0.05$). Statistical significance was lowest for the 3 bill dimensions ($F = 20.9\text{-}73.7$), reflecting the similarity in bill size and shape between males and females. The largest effect was seen in wing length ($F = 603.1$), which is relatively dimorphic sexually and which varies strongly among populations (see next chapter). Sex also had a highly significant effect

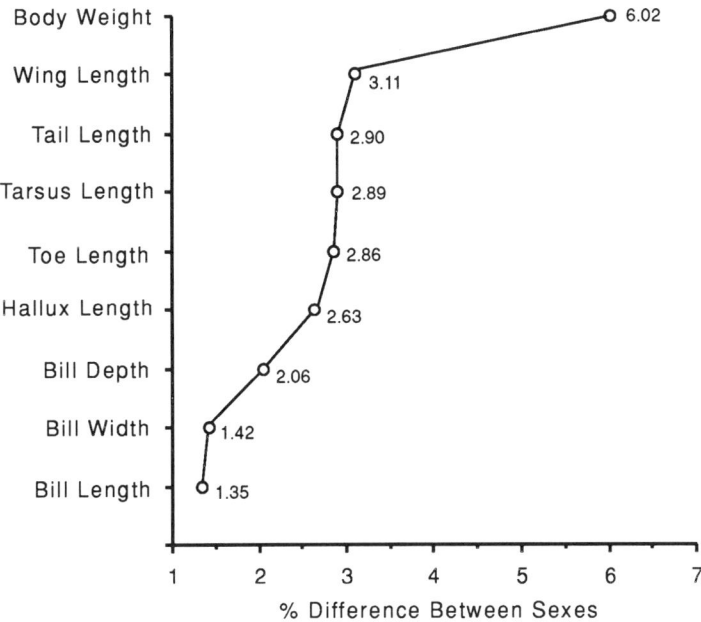

FIGURE 12. Diagram of the percentage sexual size dimorphism in *Parus inornatus* for different morphometric characters.

on tail length ($F = 257.6$) and tarsus length ($F = 281.5$), more so than for body weight ($F = 193.9$) and the two toe measurements (middle toe, $F = 165.3$; hallux, $F = 130.1$). On the basis of these results, sexes were kept separate in future analyses of size variation.

In contrast to the MANOVA results for size variation, I found no evidence of sexual dichromatism in any of the 9 parameters of color measured ($F = 0.0-2.4$, $P > 0.05$). Therefore, data from both sexes were combined in subsequent analyses of color variation.

Ontogenetic Variation

Ontogenetic differences can have a major effect on size variability within as well as between natural populations (R.S. Thorpe 1983). Furthermore, age class (i.e., juvenile versus adult) also has been reported to influence levels of intraspecific genetic variation in birds (M.A. McDonald and Smith 1990). The ontogenetic component of size variation is easier to study for mammals (e.g., Daly and Patton 1986; Patton and Smith 1990; Lara et al. 1992) than for birds, especially small songbirds, because of the relative difficulty of aging the latter on the basis of study skins. Nonetheless, several workers have developed useful aging criteria for passerine taxa (e.g., see J. Davis 1957; N.K. Johnson 1974). Pyle et al. (1987) noted, however, that no reliable plumage criteria are known for aging *Parus inornatus*. Furthermore, the extent of post-juvenal molt in this species needs further study (see previous section on molt). Accordingly, I was unable to age specimens past the fluffy juvenile period and was forced to rely on age data provided on museum tags. MANOVA was used to analyze specimens of recorded age (immature, adult) to test the effect of age, independent of sex or sample area, on size variability in *P. inornatus*. At least 50% of the specimens lacked age data and thus were omitted from the analysis. Although exclusion of these samples reduced the power of the analysis, age did not show a significant effect on size variation in any character (F = 0.0-2.3, P > 0.05). Accordingly, specimens of each sex were combined without regard to age in further analyses of size variation. Birds of different age classes did not differ in color.

GEOGRAPHIC VARIATION IN SIZE

Univariate Patterns of Size Variation

With the availability of complex computer packages for analyzing multivariate data, recent studies of geographic variation (e.g., Zink 1986; Patton and Smith 1990) have tended to exclude patterns shown by individual characters. Such data, however, can reveal information that may otherwise be hidden in multivariate analyses. Results of the sums-of-squares simultaneous test procedure (SS-STP) showed strong geographic structuring among populations of *P. inornatus* in most morphometric characters. To assess whether both sexes exhibit similar patterns of variation, population means for individual characters were regressed between males and females. Except for body weight, all characters showed a highly significant relationship (wing length, r = 0.95; tail length, r = 0.93; tarsus length, r = 0.88; length of middle toe, r = 0.84; hallux length, r = 0.84; bill length, r = 0.95; bill width, r = 0.93; bill depth, r = 0.87). Although the regression for cube root of body weight was also significant, this correlation (r = 0.52) was weaker than for the other characters. Nonetheless, the results clearly indicate that geographic trends in size are similar for the two sexes; thus, only the SS-STP data for males are presented.

Wing length (Fig. 13). Populations in the Coast, Transverse, and Peninsular ranges of California varied clinally from north to south, with wing length gradually increasing from the North Sacramento Valley (sample area 4) to the Sierra San Pedro Mártir (29) in northern Baja California. The isolated population (30) in southern Baja California was intermediate in size, being notably smaller than populations farther north. Wing length also increased clinally eastward from the San Bernardino (22) and San Jacinto (23) mountains to Joshua Tree (24) and the Eastern Mojave (37).

The smallest birds occurred on the east side of the Central Valley and the west slope of the Sierra Nevada (sample areas 6, 9, and 13-15). Wing length increased slightly

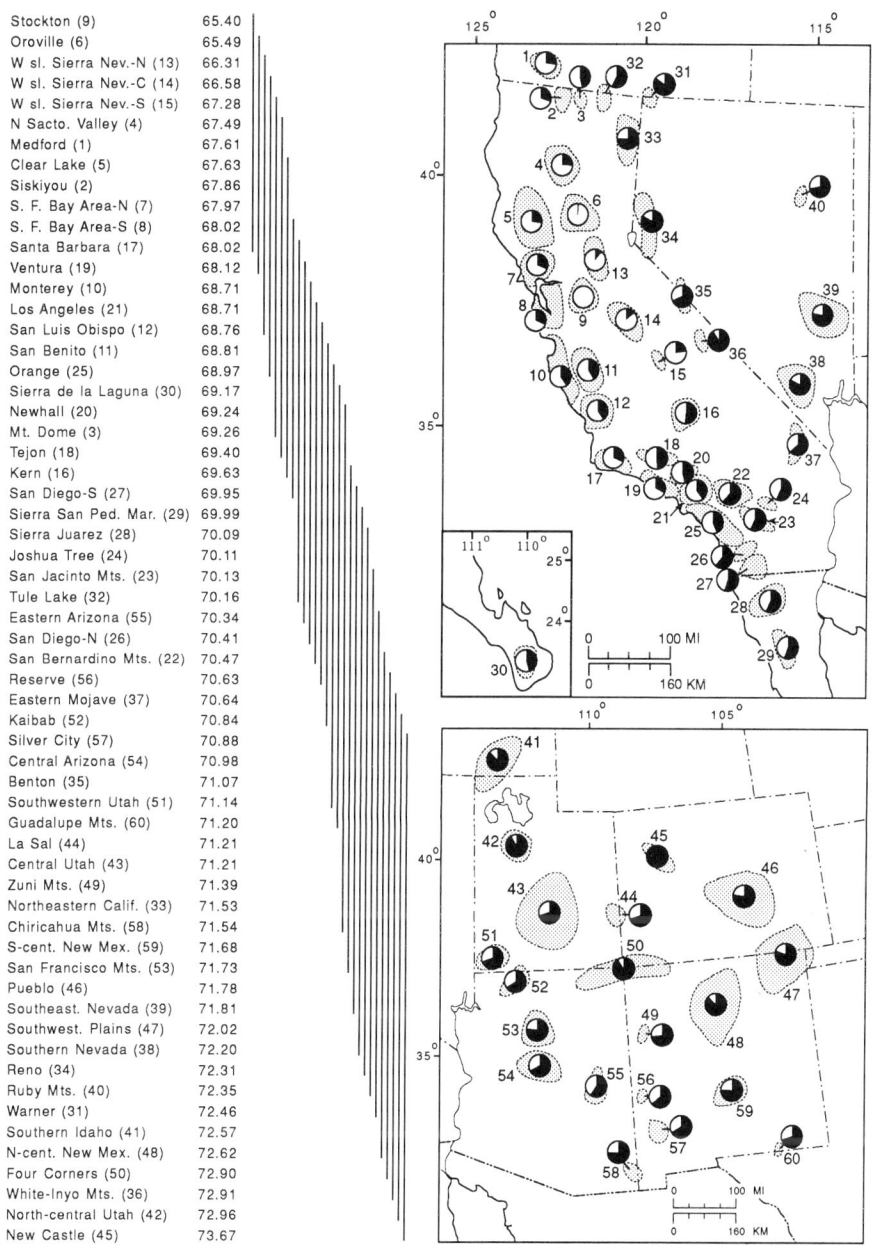

Stockton (9) 65.40
Oroville (6) 65.49
W sl. Sierra Nev.-N (13) 66.31
W sl. Sierra Nev.-C (14) 66.58
W sl. Sierra Nev.-S (15) 67.28
N Sacto. Valley (4) 67.49
Medford (1) 67.61
Clear Lake (5) 67.63
Siskiyou (2) 67.86
S. F. Bay Area-N (7) 67.97
S. F. Bay Area-S (8) 68.02
Santa Barbara (17) 68.02
Ventura (19) 68.12
Monterey (10) 68.71
Los Angeles (21) 68.71
San Luis Obispo (12) 68.76
San Benito (11) 68.81
Orange (25) 68.97
Sierra de la Laguna (30) 69.17
Newhall (20) 69.24
Mt. Dome (3) 69.26
Tejon (18) 69.40
Kern (16) 69.63
San Diego-S (27) 69.95
Sierra San Ped. Mar. (29) 69.99
Sierra Juarez (28) 70.09
Joshua Tree (24) 70.11
San Jacinto Mts. (23) 70.13
Tule Lake (32) 70.16
Eastern Arizona (55) 70.34
San Diego-N (26) 70.41
San Bernardino Mts. (22) 70.47
Reserve (56) 70.63
Eastern Mojave (37) 70.64
Kaibab (52) 70.84
Silver City (57) 70.88
Central Arizona (54) 70.98
Benton (35) 71.07
Southwestern Utah (51) 71.14
Guadalupe Mts. (60) 71.20
La Sal (44) 71.21
Central Utah (43) 71.21
Zuni Mts. (49) 71.39
Northeastern Calif. (33) 71.53
Chiricahua Mts. (58) 71.54
S-cent. New Mex. (59) 71.68
San Francisco Mts. (53) 71.73
Pueblo (46) 71.78
Southeast. Nevada (39) 71.81
Southwest. Plains (47) 72.02
Southern Nevada (38) 72.20
Reno (34) 72.31
Ruby Mts. (40) 72.35
Warner (31) 72.46
Southern Idaho (41) 72.57
N-cent. New Mex. (48) 72.62
Four Corners (50) 72.90
White-Inyo Mts. (36) 72.91
North-central Utah (42) 72.96
New Castle (45) 73.67

WING LENGTH - MALES

FIGURE 13. Results of the sums-of-squares simultaneous test procedure (SS-STP) applied to mean wing length for 60 samples of the *Parus inornatus* complex (males). Sample areas (stippled) are numbered as in Figs. 3-4 and Table 2, and are listed in increasing order of means. Adjacent vertical lines denote samples whose means are not significantly different from each other ($P < 0.05$). Pie diagrams illustrate the patterns of variation geographically, whereby means are scaled from the smallest (0%, open pie) to the largest (100%, closed pie) value. Because the subsets are formed independent of geography, samples from widely disparate geographic areas may be grouped together if their means are not significantly different.

from north to south among these populations, then rose sharply at the southern end of the Sierra Nevada (Kern, sample area 16). Males from Kern were essentially identical to those from Tejon (18) in the Transverse Range. The small size of titmice from the western Sierra Nevada contrasted strikingly with the larger birds found along the California-Nevada border. In northern California and southern Oregon, a fairly sharp cline separated samples along an east-west transect from Medford (1) to Warner (31).

Interior samples showed a greater mosaic of patterns in wing length compared to those from the Pacific slope. Nonetheless, titmice from the more southerly interior populations (i.e., sample areas 49 and 51-60) were clearly smaller than birds farther north. Males from the Southwest were slightly larger than those from southern and southeastern California, although the differences were not always statistically significant. Wing length was substantially greater in North-central New Mexico (48) and Four Corners (50) than in adjacent populations, resembling that of birds from Southern Idaho (41) or North-central Utah (42). Males with the longest wings were from New Castle, Colorado (45).

Tail length (Fig. 14). Tail length showed the same general patterns as wing length, although differences were sharper among certain groups of populations. In parallel with wing length, tail length also increased clinally from the North Sacramento Valley (4) to the Sierra San Pedro Mártir (29). However, tail length exhibited a steeper cline than wing length, particularly toward the south. Birds from these southern coastal populations also had much longer tails than those from the Sierra de la Laguna (30) in southern Baja California. As with wing length, the shortest-tailed birds occurred on the east side of the Central Valley and west slope of the Sierra Nevada (sample areas 6, 9, 13-15). Birds from these samples had significantly shorter tails than those along the western edge of the Great Basin (sample areas 31, 33-36). Males from Kern (16) had intermediate tail lengths compared to samples from the southern Sierra Nevada (15) and Tejon (18).

Tail length, like wing length, increased sharply across a fairly narrow gradient in northern California and southern Oregon. Although birds from Medford (1) and Siskiyou (2) did not differ in wing length, those from Siskiyou had notably longer tails. Tail length increased again between Siskiyou and Mt. Dome (3), and then showed a slight decline eastward to Tule Lake (32). Birds from Tule Lake had shorter tails than those from Warner (31) or Northeastern California (33).

In contrast to wing length, tail length decreased from the San Bernardino (22) and San Jacinto (23) mountains in southern California to the Eastern Mojave (37). From there, tail length increased northward to reach an extreme in the Ruby Mountains (40). Birds with the longest tails occurred in the most northern interior samples, including Warner (31), Ruby Mountains (40), Southern Idaho (41), North-central Utah (42), and New Castle (45). Tail length declined again toward the Southwest, where titmice had relatively short tails compared to those from the northern and western Great Basin and from southern California and northern Baja California. Within the Southwest, birds from Central (54) and Eastern (55) Arizona were distinguished by the shortest tails.

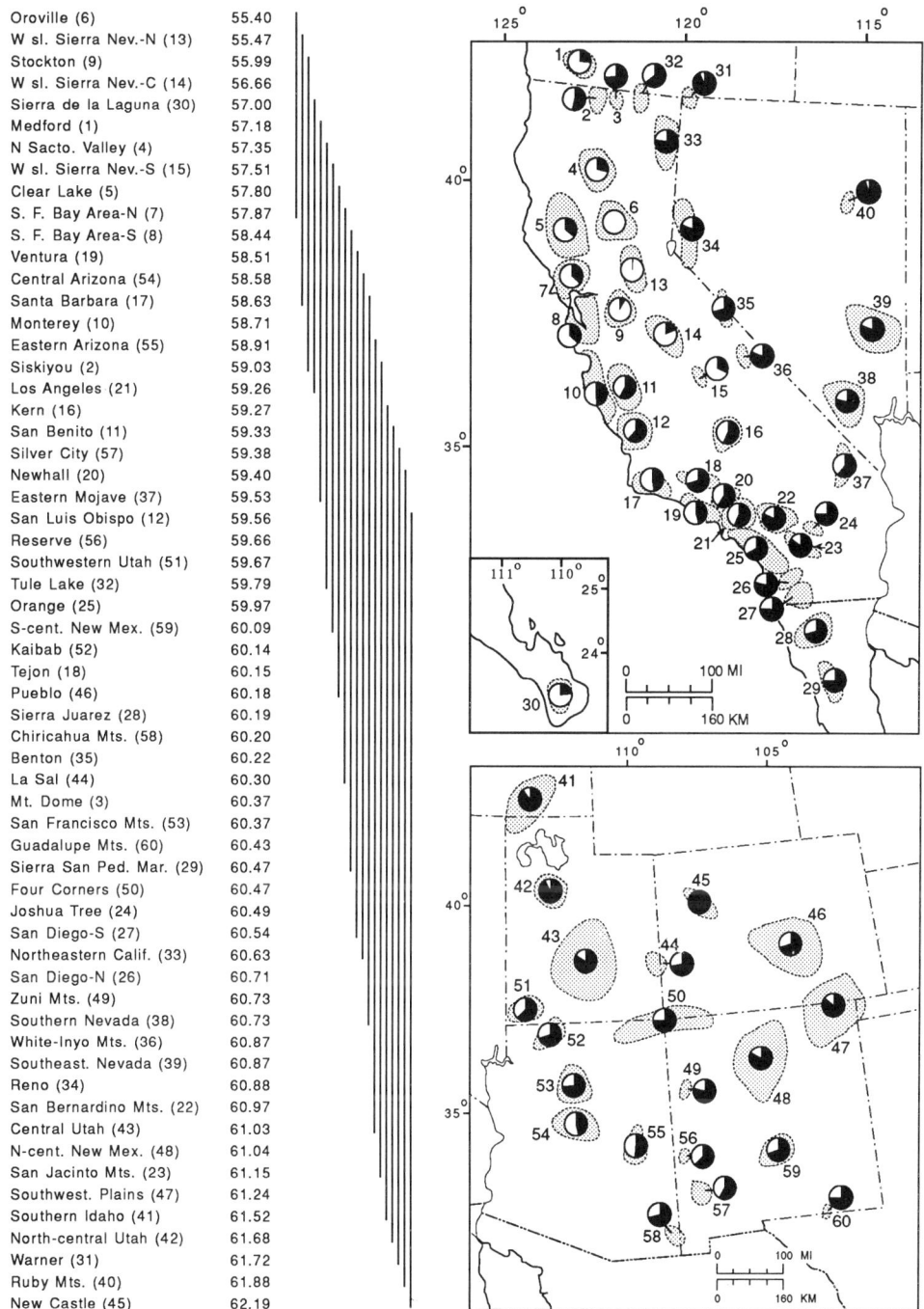

Oroville (6)	55.40
W sl. Sierra Nev.-N (13)	55.47
Stockton (9)	55.99
W sl. Sierra Nev.-C (14)	56.66
Sierra de la Laguna (30)	57.00
Medford (1)	57.18
N Sacto. Valley (4)	57.35
W sl. Sierra Nev.-S (15)	57.51
Clear Lake (5)	57.80
S. F. Bay Area-N (7)	57.87
S. F. Bay Area-S (8)	58.44
Ventura (19)	58.51
Central Arizona (54)	58.58
Santa Barbara (17)	58.63
Monterey (10)	58.71
Eastern Arizona (55)	58.91
Siskiyou (2)	59.03
Los Angeles (21)	59.26
Kern (16)	59.27
San Benito (11)	59.33
Silver City (57)	59.38
Newhall (20)	59.40
Eastern Mojave (37)	59.53
San Luis Obispo (12)	59.56
Reserve (56)	59.66
Southwestern Utah (51)	59.67
Tule Lake (32)	59.79
Orange (25)	59.97
S-cent. New Mex. (59)	60.09
Kaibab (52)	60.14
Tejon (18)	60.15
Pueblo (46)	60.18
Sierra Juarez (28)	60.19
Chiricahua Mts. (58)	60.20
Benton (35)	60.22
La Sal (44)	60.30
Mt. Dome (3)	60.37
San Francisco Mts. (53)	60.37
Guadalupe Mts. (60)	60.43
Sierra San Ped. Mar. (29)	60.47
Four Corners (50)	60.47
Joshua Tree (24)	60.49
San Diego-S (27)	60.54
Northeastern Calif. (33)	60.63
San Diego-N (26)	60.71
Zuni Mts. (49)	60.73
Southern Nevada (38)	60.73
White-Inyo Mts. (36)	60.87
Southeast. Nevada (39)	60.87
Reno (34)	60.88
San Bernardino Mts. (22)	60.97
Central Utah (43)	61.03
N-cent. New Mex. (48)	61.04
San Jacinto Mts. (23)	61.15
Southwest. Plains (47)	61.24
Southern Idaho (41)	61.52
North-central Utah (42)	61.68
Warner (31)	61.72
Ruby Mts. (40)	61.88
New Castle (45)	62.19

TAIL LENGTH - MALES

FIGURE 14. SS-STP applied to mean tail length for 60 samples of the *Parus inornatus* complex (males). The analysis and presentation of geographic variation in this character is the same as for wing length (see legend to Fig. 13).

Tarsus length (Fig. 15). Patterns of variation in tarsus length generally followed those exhibited by wing and/or tail measurements. Birds with the shortest tarsi occurred in southwestern Oregon (1), northwestern California (2-3), on the east side of the Central Valley (6, 9, 13-15), and in the Sierra de la Laguna (30). Tarsus length increased clinally southward from the North Sacramento Valley (4) to the Sierra San Pedro Mártir (29); birds from southern California and northern Baja California had longer tarsi compared to other coastal and some interior samples. An abrupt increase in tarsus length was observed between Mt. Dome (3) and Tule Lake (32) in northern California. Tarsus length also increased sharply across the Sierra Nevada from the Pacific slope to the interior. In general, tarsus length was reduced in samples from the Southwest (49, 51-60). However, a mosaic pattern of variation was evident again in the interior. The longest tarsi were measured from birds in the Four Corners area (50), which was surrounded by intermediate samples. In keeping with their larger size, birds from the more northern interior samples (31, 33, 40, 41, 42, 45) had relatively long tarsi.

Toe dimensions (Fig. 16). Both the middle toe and hallux exhibited similar patterns of geographic variation; thus, only the SS-STP results for length of middle toe are shown. Titmice with the longest toes were found in southern California, particularly in the San Bernardino (22), Orange (25), and San Diego (26-27) sample areas. Toe length declined southward to Baja California (sample areas 28-30), eastward to the Mojave Desert (sample areas 24 and 37-38), and northward along the Coast Ranges to the North Sacramento Valley (4). The relatively small titmice that inhabit the eastern Sacramento Valley and west slope of the Sierra Nevada (sample areas 6, 9, 13-15) likewise had short toes compared to other Pacific slope samples. Toe length in the Kern (16) sample was similar to that in the southern Sierra Nevada (15), and much shorter than the nearest coastal sample in the Transverse Range (18). Although birds from the western Great Basin (33-36) generally had longer middle toes than those from the west slope of the Sierra Nevada, hallux length was shorter or equivalent. Titmice from Warner (31) had significantly longer toes compared to those from Tule Lake (32), which in turn had slightly shorter toes relative to samples farther west (sample areas 1-3).

Like other size characters, length of the middle toe and hallux exhibited a mosaic pattern throughout the interior. Nonetheless, both toe measurements showed a tendency toward reduced size in interior populations compared to coastal populations. This finding is especially interesting in view of the larger overall size of titmice from the interior relative to the Pacific slope.

Bill dimensions (Figs. 17-19). Bill characters showed strong geographic structuring in both size and shape components. Bill length (Fig. 17) was smallest in northern and central samples of the Pacific slope (1-20) and in the Sierra de la Laguna (30). By comparison, bills were moderately longer in birds from southern California (21-27) and northern Baja California (28-29). Although bill length did not differ between Joshua Tree (24) and other coastal samples to the west, titmice from Joshua Tree had significantly shorter bills compared to those from the Eastern Mojave (37). Another major shift in bill

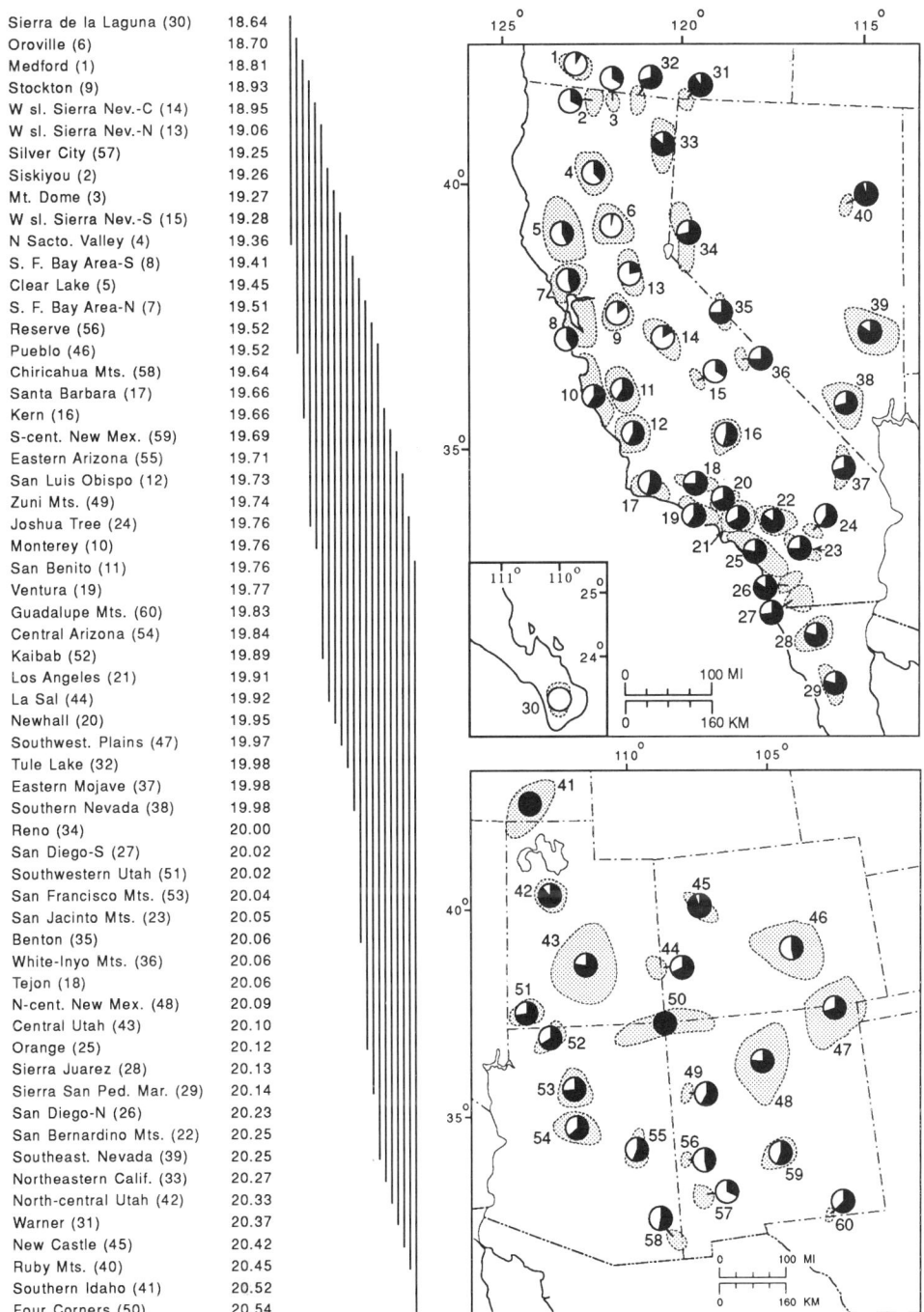

Sample	Value
Sierra de la Laguna (30)	18.64
Oroville (6)	18.70
Medford (1)	18.81
Stockton (9)	18.93
W sl. Sierra Nev.-C (14)	18.95
W sl. Sierra Nev.-N (13)	19.06
Silver City (57)	19.25
Siskiyou (2)	19.26
Mt. Dome (3)	19.27
W sl. Sierra Nev.-S (15)	19.28
N Sacto. Valley (4)	19.36
S. F. Bay Area-S (8)	19.41
Clear Lake (5)	19.45
S. F. Bay Area-N (7)	19.51
Reserve (56)	19.52
Pueblo (46)	19.52
Chiricahua Mts. (58)	19.64
Santa Barbara (17)	19.66
Kern (16)	19.66
S-cent. New Mex. (59)	19.69
Eastern Arizona (55)	19.71
San Luis Obispo (12)	19.73
Zuni Mts. (49)	19.74
Joshua Tree (24)	19.76
Monterey (10)	19.76
San Benito (11)	19.76
Ventura (19)	19.77
Guadalupe Mts. (60)	19.83
Central Arizona (54)	19.84
Kaibab (52)	19.89
Los Angeles (21)	19.91
La Sal (44)	19.92
Newhall (20)	19.95
Southwest. Plains (47)	19.97
Tule Lake (32)	19.98
Eastern Mojave (37)	19.98
Southern Nevada (38)	19.98
Reno (34)	20.00
San Diego-S (27)	20.02
Southwestern Utah (51)	20.02
San Francisco Mts. (53)	20.04
San Jacinto Mts. (23)	20.05
Benton (35)	20.06
White-Inyo Mts. (36)	20.06
Tejon (18)	20.06
N-cent. New Mex. (48)	20.09
Central Utah (43)	20.10
Orange (25)	20.12
Sierra Juarez (28)	20.13
Sierra San Ped. Mar. (29)	20.14
San Diego-N (26)	20.23
San Bernardino Mts. (22)	20.25
Southeast. Nevada (39)	20.25
Northeastern Calif. (33)	20.27
North-central Utah (42)	20.33
Warner (31)	20.37
New Castle (45)	20.42
Ruby Mts. (40)	20.45
Southern Idaho (41)	20.52
Four Corners (50)	20.54

TARSUS LENGTH - MALES

FIGURE 15. SS-STP applied to mean tarsus length for 60 samples of the *Parus inornatus* complex (males). The analysis and presentation of geographic variation in this character is the same as for wing length (see legend to Fig. 13).

Pueblo (46)	13.34
Guadalupe Mts. (60)	13.43
Silver City (57)	13.44
Southwestern Utah (51)	13.52
Southwest. Plains (47)	13.54
La Sal (44)	13.55
Oroville (6)	13.61
Reserve (56)	13.62
W sl. Sierra Nev.-C (14)	13.62
Eastern Arizona (55)	13.62
Central Utah (43)	13.63
N-cent. New Mex. (48)	13.65
Chiricahua Mts. (58)	13.69
Central Arizona (54)	13.69
Southern Nevada (38)	13.70
S-cent. New Mex. (59)	13.71
Tule Lake (32)	13.75
Eastern Mojave (37)	13.76
Kern (16)	13.79
W sl. Sierra Nev.-S (15)	13.82
W sl. Sierra Nev.-N (13)	13.82
Southern Idaho (41)	13.84
Medford (1)	13.88
Reno (34)	13.89
Four Corners (50)	13.89
Zuni Mts. (49)	13.91
Northeastern Calif. (33)	13.91
Siskiyou (2)	13.92
Mt. Dome (3)	13.93
Kaibab (52)	13.97
Stockton (9)	13.97
North-central Utah (42)	13.99
White-Inyo Mts. (36)	14.00
Sierra de la Laguna (30)	14.10
Clear Lake (5)	14.11
New Castle (45)	14.12
Ruby Mts. (40)	14.14
San Francisco Mts. (53)	14.14
S. F. Bay Area-S (8)	14.14
S. F. Bay Area-N (7)	14.15
N Sacto. Valley (4)	14.16
Sierra San Ped. Mar. (29)	14.19
Southeast. Nevada (39)	14.20
Benton (35)	14.21
Los Angeles (21)	14.22
San Jacinto Mts. (23)	14.23
Joshua Tree (24)	14.26
San Luis Obispo (12)	14.26
Sierra Juarez (28)	14.27
Warner (31)	14.28
Tejon (18)	14.36
Santa Barbara (17)	14.38
Newhall (20)	14.38
San Benito (11)	14.40
Monterey (10)	14.41
Ventura (19)	14.44
San Diego-S (27)	14.50
San Diego-N (26)	14.52
San Bernardino Mts. (22)	14.59
Orange (25)	14.67

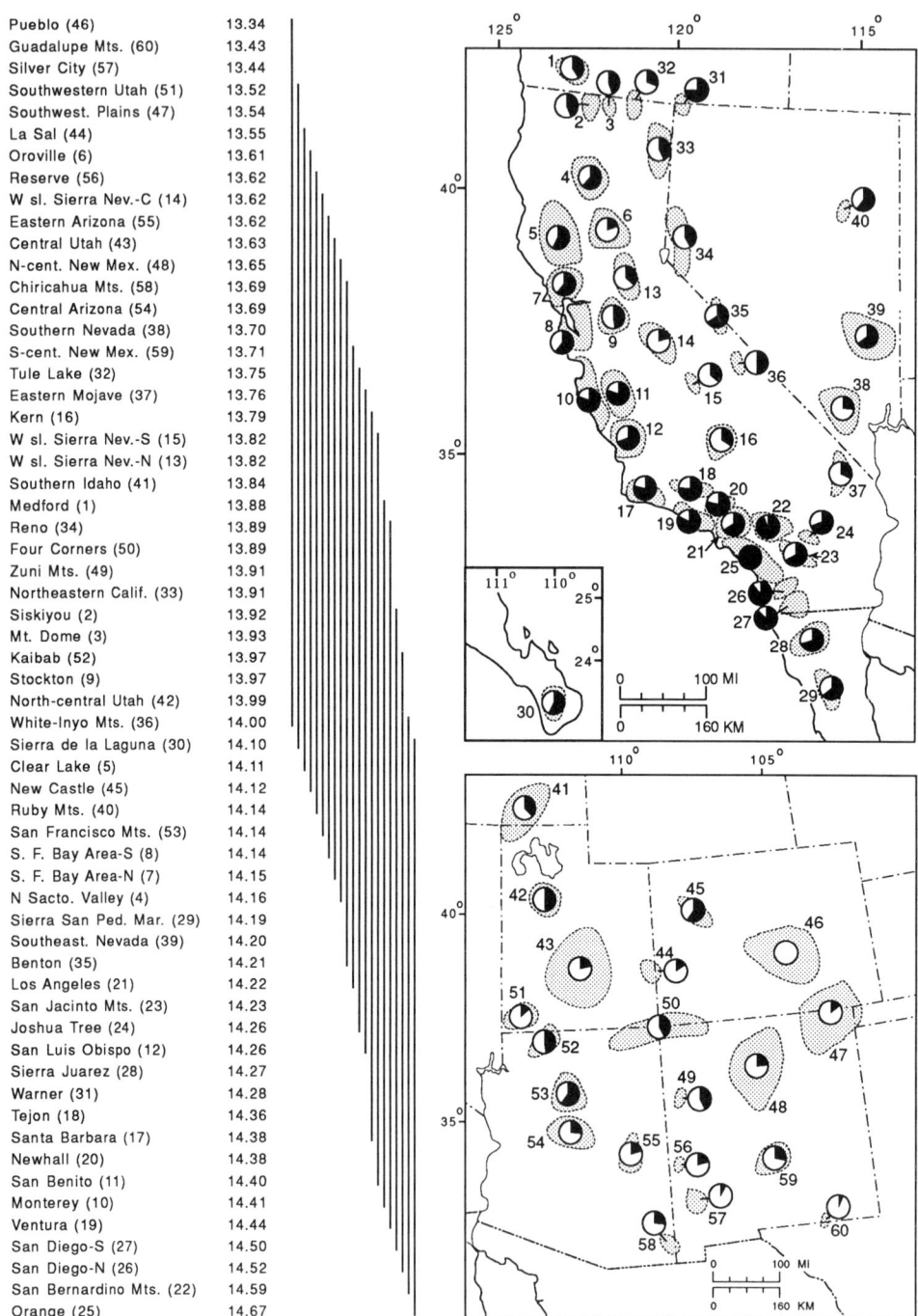

TOE LENGTH - MALES

FIGURE 16. SS-STP applied to mean length of the middle toe for 60 samples of the *Parus inornatus* complex (males). The analysis and presentation of geographic variation in this character is the same as for wing length (see legend to Fig. 13).

length occurred in northern California between Mt. Dome (3) and Tule Lake (32). The increase in bill length toward the interior generally characterized populations from the Great Basin to the Rocky Mountains and Southwest. Within this broad region, however, bill length declined from north and west to south and east. Birds from the more southerly samples in Arizona (55, 58) and New Mexico (56-57, 59) had bills of medium length that did not differ significantly from those in southern California and northern Baja California.

Bill depth (Fig. 18) and width (Fig. 19) showed a steeper cline than bill length in samples from southern California and northern Baja California, while birds from the Sierra de la Laguna (30) again were substantially smaller in these dimensions. Titmice from Santa Barbara (17) to Newhall (20) were intermediate in terms of bill depth between samples to the north and south, as were those from Santa Barbara and Tejon (18) in terms of bill width. Bill depth and width did not differ significantly between the southern coastal samples and Joshua Tree (24). However, birds from Joshua Tree had deeper and significantly narrower bills compared to those from the Eastern Mojave (37).

The depth and width of bills generally increased toward the northern Great Basin in concordance with an increase in bill length. Thus, individuals from Warner (31) had a much larger and heavier bill relative to samples farther west. In eastern California and western Nevada (34-36), bill depth increased less than either length or width. From Warner, bill depth and width increased eastward to the Ruby Mountains (40), Southern Idaho (41), and North-central Utah (42). Birds with deep, wide bills also occurred at Four Corners (50) and in the Guadalupe Mountains (60), although sample sizes were small. Populations from Arizona and New Mexico generally resembled those from southern California and northern Baja in one or both dimensions. However, except for the Guadalupe Mountains, titmice from the more southerly interior samples (55-59) differed in having bills of lesser depth.

Cube root of body weight (Fig. 20). Of all size characters, cube root of body weight showed the most chaotic pattern of variation. Although males from the northern Great Basin tended to be slightly larger on average, no clear trends were observed. High variances and small sample sizes may account partially for the lack of statistical significance. Sample sizes for body weight generally were smaller than for other characters because many specimens lacked weight data. Furthermore, given that specimens were examined regardless of the date of collection, seasonal variability in body weight may have confounded the analysis. Nonetheless, there may be some real component to the lack of geographic structure in the data. Grinnell (1923:137) found no difference in body mass between 39 specimens of *P. i. inornatus* and 19 specimens of the interior form "*griseus*" despite a "relatively longer tail, longer wings, and larger bill" in "*griseus*."

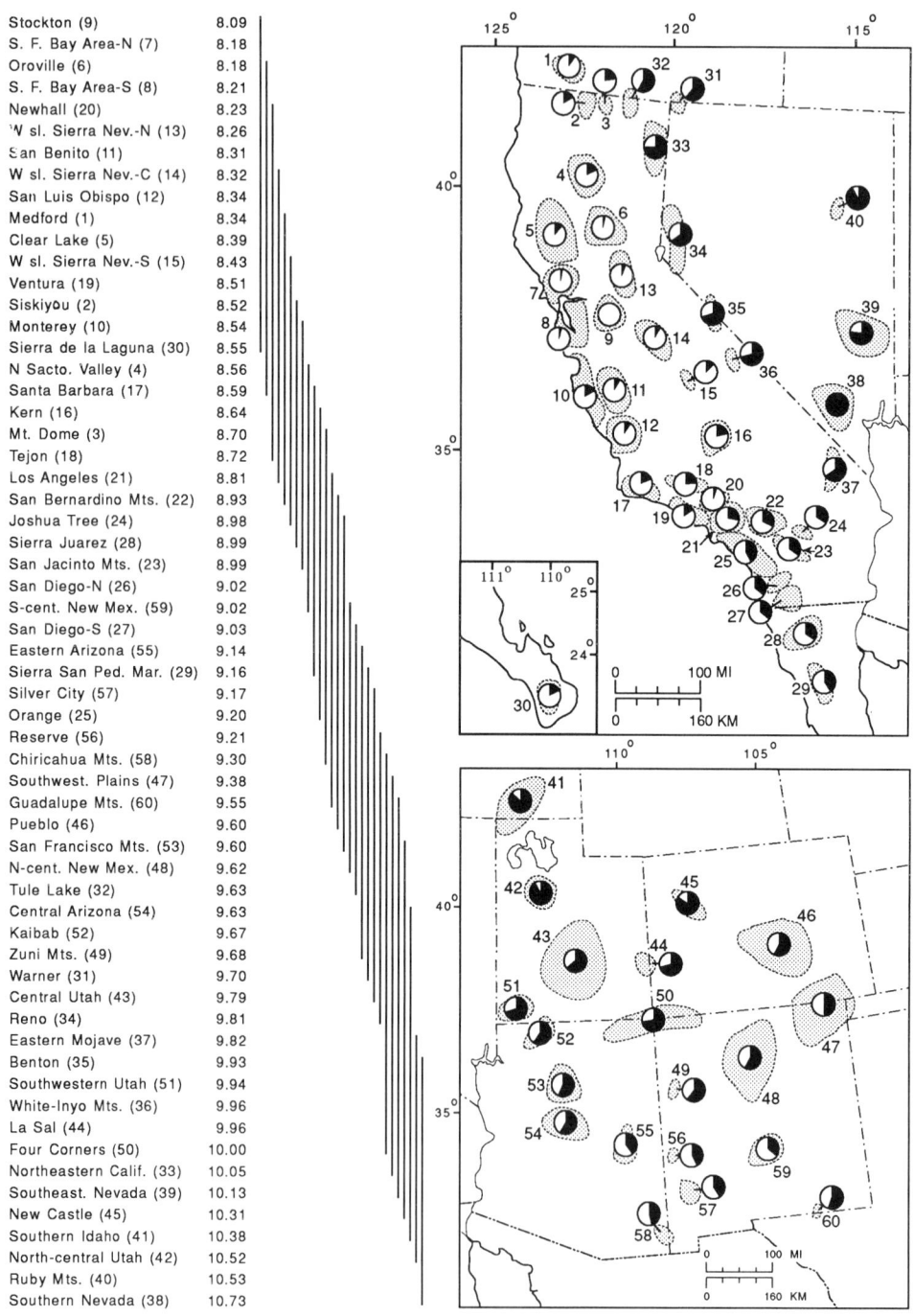

Stockton (9)	8.09
S. F. Bay Area-N (7)	8.18
Oroville (6)	8.18
S. F. Bay Area-S (8)	8.21
Newhall (20)	8.23
W sl. Sierra Nev.-N (13)	8.26
San Benito (11)	8.31
W sl. Sierra Nev.-C (14)	8.32
San Luis Obispo (12)	8.34
Medford (1)	8.34
Clear Lake (5)	8.39
W sl. Sierra Nev.-S (15)	8.43
Ventura (19)	8.51
Siskiyou (2)	8.52
Monterey (10)	8.54
Sierra de la Laguna (30)	8.55
N Sacto. Valley (4)	8.56
Santa Barbara (17)	8.59
Kern (16)	8.64
Mt. Dome (3)	8.70
Tejon (18)	8.72
Los Angeles (21)	8.81
San Bernardino Mts. (22)	8.93
Joshua Tree (24)	8.98
Sierra Juarez (28)	8.99
San Jacinto Mts. (23)	8.99
San Diego-N (26)	9.02
S-cent. New Mex. (59)	9.02
San Diego-S (27)	9.03
Eastern Arizona (55)	9.14
Sierra San Ped. Mar. (29)	9.16
Silver City (57)	9.17
Orange (25)	9.20
Reserve (56)	9.21
Chiricahua Mts. (58)	9.30
Southwest. Plains (47)	9.38
Guadalupe Mts. (60)	9.55
Pueblo (46)	9.60
San Francisco Mts. (53)	9.60
N-cent. New Mex. (48)	9.62
Tule Lake (32)	9.63
Central Arizona (54)	9.63
Kaibab (52)	9.67
Zuni Mts. (49)	9.68
Warner (31)	9.70
Central Utah (43)	9.79
Reno (34)	9.81
Eastern Mojave (37)	9.82
Benton (35)	9.93
Southwestern Utah (51)	9.94
White-Inyo Mts. (36)	9.96
La Sal (44)	9.96
Four Corners (50)	10.00
Northeastern Calif. (33)	10.05
Southeast. Nevada (39)	10.13
New Castle (45)	10.31
Southern Idaho (41)	10.38
North-central Utah (42)	10.52
Ruby Mts. (40)	10.53
Southern Nevada (38)	10.73

BILL LENGTH - MALES

FIGURE 17. SS-STP applied to mean bill length for 60 samples of the *Parus inornatus* complex (males). The analysis and presentation of geographic variation in this character is the same as for wing length (see legend to Fig. 13).

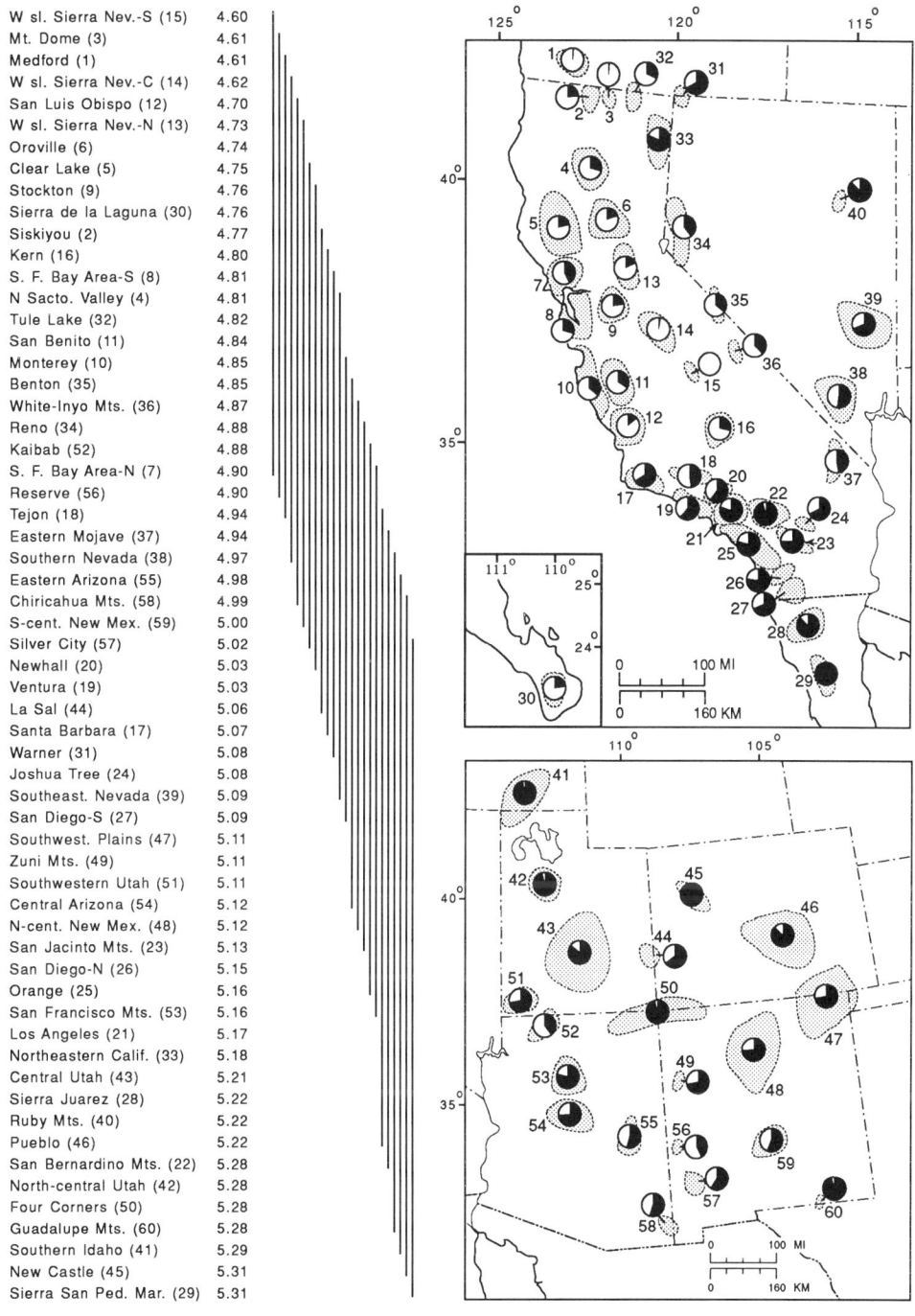

W sl. Sierra Nev.-S (15)	4.60
Mt. Dome (3)	4.61
Medford (1)	4.61
W sl. Sierra Nev.-C (14)	4.62
San Luis Obispo (12)	4.70
W sl. Sierra Nev.-N (13)	4.73
Oroville (6)	4.74
Clear Lake (5)	4.75
Stockton (9)	4.76
Sierra de la Laguna (30)	4.76
Siskiyou (2)	4.77
Kern (16)	4.80
S. F. Bay Area-S (8)	4.81
N Sacto. Valley (4)	4.81
Tule Lake (32)	4.82
San Benito (11)	4.84
Monterey (10)	4.85
Benton (35)	4.85
White-Inyo Mts. (36)	4.87
Reno (34)	4.88
Kaibab (52)	4.88
S. F. Bay Area-N (7)	4.90
Reserve (56)	4.90
Tejon (18)	4.94
Eastern Mojave (37)	4.94
Southern Nevada (38)	4.97
Eastern Arizona (55)	4.98
Chiricahua Mts. (58)	4.99
S-cent. New Mex. (59)	5.00
Silver City (57)	5.02
Newhall (20)	5.03
Ventura (19)	5.03
La Sal (44)	5.06
Santa Barbara (17)	5.07
Warner (31)	5.08
Joshua Tree (24)	5.08
Southeast. Nevada (39)	5.09
San Diego-S (27)	5.09
Southwest. Plains (47)	5.11
Zuni Mts. (49)	5.11
Southwestern Utah (51)	5.11
Central Arizona (54)	5.12
N-cent. New Mex. (48)	5.12
San Jacinto Mts. (23)	5.13
San Diego-N (26)	5.15
Orange (25)	5.16
San Francisco Mts. (53)	5.16
Los Angeles (21)	5.17
Northeastern Calif. (33)	5.18
Central Utah (43)	5.21
Sierra Juarez (28)	5.22
Ruby Mts. (40)	5.22
Pueblo (46)	5.22
San Bernardino Mts. (22)	5.28
North-central Utah (42)	5.28
Four Corners (50)	5.28
Guadalupe Mts. (60)	5.28
Southern Idaho (41)	5.29
New Castle (45)	5.31
Sierra San Ped. Mar. (29)	5.31

BILL DEPTH - MALES

FIGURE 18. SS-STP applied to mean bill depth for 60 samples of the *Parus inornatus* complex (males). The analysis and presentation of geographic variation in this character is the same as for wing length (see legend to Fig. 13).

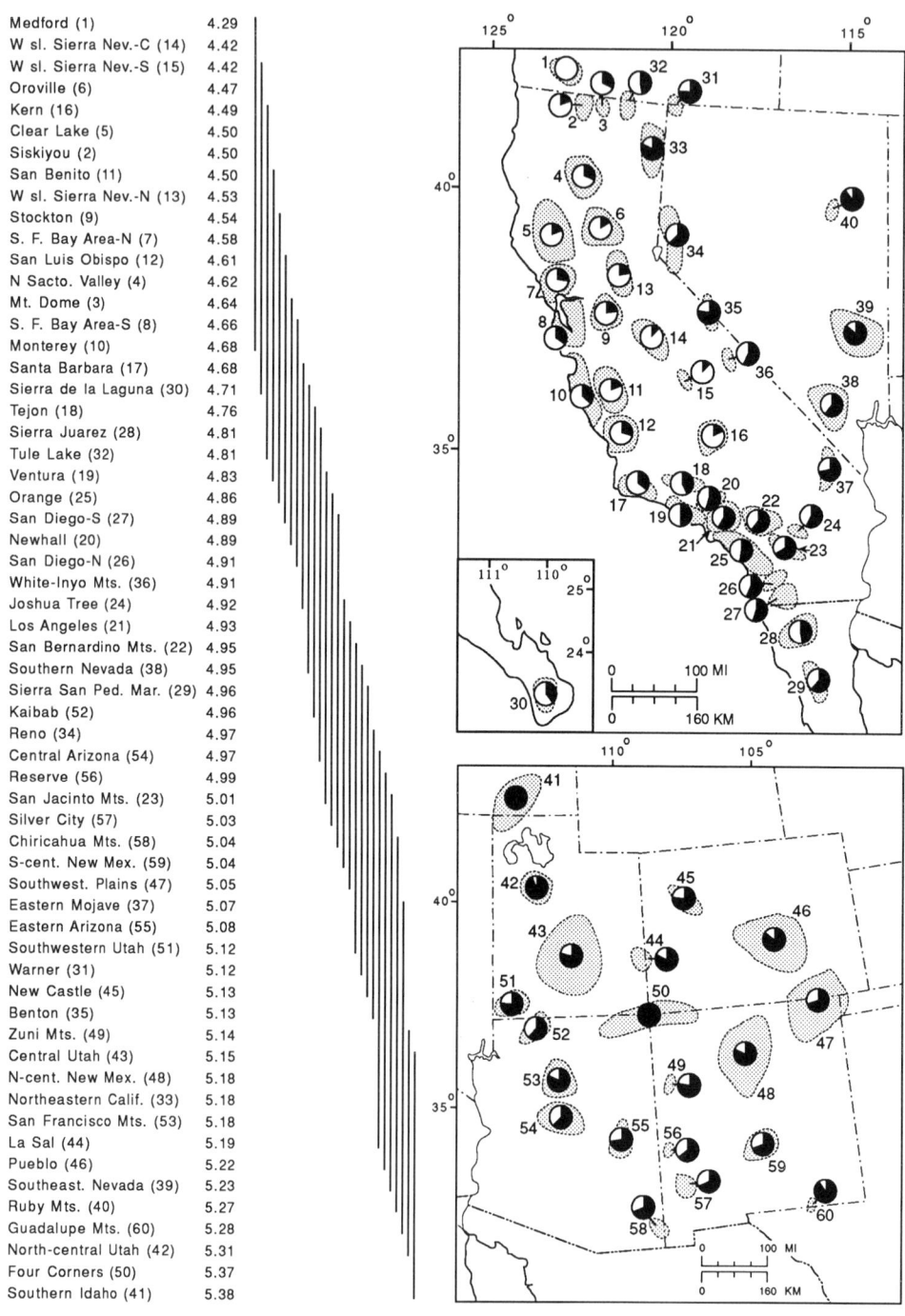

Medford (1)	4.29
W sl. Sierra Nev.-C (14)	4.42
W sl. Sierra Nev.-S (15)	4.42
Oroville (6)	4.47
Kern (16)	4.49
Clear Lake (5)	4.50
Siskiyou (2)	4.50
San Benito (11)	4.50
W sl. Sierra Nev.-N (13)	4.53
Stockton (9)	4.54
S. F. Bay Area-N (7)	4.58
San Luis Obispo (12)	4.61
N Sacto. Valley (4)	4.62
Mt. Dome (3)	4.64
S. F. Bay Area-S (8)	4.66
Monterey (10)	4.68
Santa Barbara (17)	4.68
Sierra de la Laguna (30)	4.71
Tejon (18)	4.76
Sierra Juarez (28)	4.81
Tule Lake (32)	4.81
Ventura (19)	4.83
Orange (25)	4.86
San Diego-S (27)	4.89
Newhall (20)	4.89
San Diego-N (26)	4.91
White-Inyo Mts. (36)	4.91
Joshua Tree (24)	4.92
Los Angeles (21)	4.93
San Bernardino Mts. (22)	4.95
Southern Nevada (38)	4.95
Sierra San Ped. Mar. (29)	4.96
Kaibab (52)	4.96
Reno (34)	4.97
Central Arizona (54)	4.97
Reserve (56)	4.99
San Jacinto Mts. (23)	5.01
Silver City (57)	5.03
Chiricahua Mts. (58)	5.04
S-cent. New Mex. (59)	5.04
Southwest. Plains (47)	5.05
Eastern Mojave (37)	5.07
Eastern Arizona (55)	5.08
Southwestern Utah (51)	5.12
Warner (31)	5.12
New Castle (45)	5.13
Benton (35)	5.13
Zuni Mts. (49)	5.14
Central Utah (43)	5.15
N-cent. New Mex. (48)	5.18
Northeastern Calif. (33)	5.18
San Francisco Mts. (53)	5.18
La Sal (44)	5.19
Pueblo (46)	5.22
Southeast. Nevada (39)	5.23
Ruby Mts. (40)	5.27
Guadalupe Mts. (60)	5.28
North-central Utah (42)	5.31
Four Corners (50)	5.37
Southern Idaho (41)	5.38

BILL WIDTH - MALES

FIGURE 19. SS-STP applied to mean bill width for 60 samples of the *Parus inornatus* complex (males). The analysis and presentation of geographic variation in this character is the same as for wing length (see legend to Fig. 13).

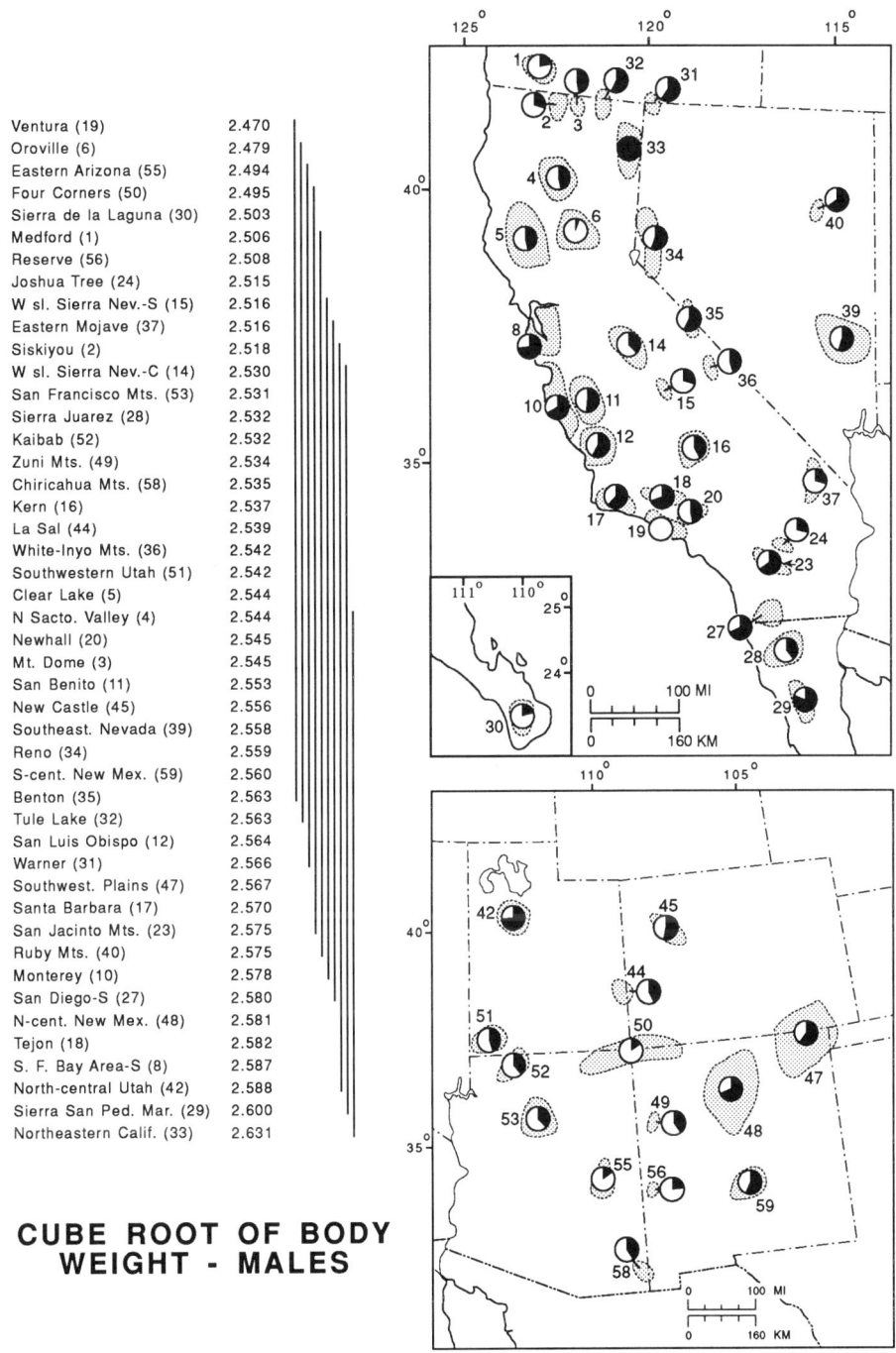

Ventura (19)	2.470
Oroville (6)	2.479
Eastern Arizona (55)	2.494
Four Corners (50)	2.495
Sierra de la Laguna (30)	2.503
Medford (1)	2.506
Reserve (56)	2.508
Joshua Tree (24)	2.515
W sl. Sierra Nev.-S (15)	2.516
Eastern Mojave (37)	2.516
Siskiyou (2)	2.518
W sl. Sierra Nev.-C (14)	2.530
San Francisco Mts. (53)	2.531
Sierra Juarez (28)	2.532
Kaibab (52)	2.532
Zuni Mts. (49)	2.534
Chiricahua Mts. (58)	2.535
Kern (16)	2.537
La Sal (44)	2.539
White-Inyo Mts. (36)	2.542
Southwestern Utah (51)	2.542
Clear Lake (5)	2.544
N Sacto. Valley (4)	2.544
Newhall (20)	2.545
Mt. Dome (3)	2.545
San Benito (11)	2.553
New Castle (45)	2.556
Southeast. Nevada (39)	2.558
Reno (34)	2.559
S-cent. New Mex. (59)	2.560
Benton (35)	2.563
Tule Lake (32)	2.563
San Luis Obispo (12)	2.564
Warner (31)	2.566
Southwest. Plains (47)	2.567
Santa Barbara (17)	2.570
San Jacinto Mts. (23)	2.575
Ruby Mts. (40)	2.575
Monterey (10)	2.578
San Diego-S (27)	2.580
N-cent. New Mex. (48)	2.581
Tejon (18)	2.582
S. F. Bay Area-S (8)	2.587
North-central Utah (42)	2.588
Sierra San Ped. Mar. (29)	2.600
Northeastern Calif. (33)	2.631

CUBE ROOT OF BODY WEIGHT - MALES

FIGURE 20. SS-STP applied to mean body weight (cube root) for 46 samples of the *Parus inornatus* complex (males). The analysis and presentation of geographic variation in this character is the same as for wing length (see legend to Fig. 13). Weight data were missing for specimens from the other 14 sample areas.

Multivariate Patterns of Size Variation

Ordination analysis. The first three principal-component axes for males and females explained 89.1% and 88.2% of the variance, respectively (Table 9). Both the magnitude and direction of the factor loadings were similar between sexes. Loadings on PC 1 indicated a general size axis based on variation in wing length, tail length, tarsus length, and the three bill dimensions. Toe measurements loaded heavily on PC 2, while the third axis primarily showed differences in body weight. The distinct patterns exhibited by overall size and toe length also were evident in the SS-STP diagrams.

TABLE 9. Character loadings on the first three principal components of variation in 9 linear measurements of size in the *Parus inornatus* complex[a]. The percentage of total variance (males=89.09%, females=88.24%) explained by each axis is indicated.

	Males			Females		
	PC 1	PC 2	PC 3	PC 1	PC 2	PC 3
Wing Length	0.936	-0.181	0.040	0.903	-0.295	-0.030
Tail Length	0.920	0.159	-0.050	0.895	0.049	-0.194
Tarsus Length	0.888	0.353	0.038	0.846	0.363	0.048
Length of Middle Toe	-0.056	0.929	0.039	0.163	0.924	-0.012
Hallux Length	-0.149	0.907	0.260	0.069	0.926	0.209
Bill Depth	0.825	0.191	0.241	0.818	0.159	0.367
Bill Width	0.923	-0.172	0.174	0.900	-0.253	0.223
Bill Length	0.890	-0.293	0.023	0.856	-0.362	0.037
Body Weight (cube root)	0.456	0.350	-0.800	0.470	0.229	-0.819
Variance Explained	56.34%	23.82%	8.93%	53.30%	24.49%	10.46%

[a] Loadings are based on a variance-covariance matrix of mean values (standardized by variance) for 60 sample areas. Weight data for males were missing from 14 sample areas; weight data for females were missing from 18 sample areas.

Multidimensional scaling analysis revealed three major groups on the first two axes for both males (Fig. 21A) and females (Fig. 22A): (I) relatively small birds with short to medium toes, which occur from Medford (1) to Kern (16) and in the Sierra de la Laguna (30) (subspecies *sequestratus*, *inornatus*, *kernensis*, and *cineraceus*); (II) birds of intermediate size but relatively long toes from Santa Barbara (17) to the Sierra San Pedro Mártir (29) (*transpositus*, *mohavensis*, and *affabilis*); and (III) moderately large to large birds, with short to medium-length toes, from the Great Basin to the Rocky Mountains and Southwest (sample areas 31-60; *zaleptus*, *ridgwayi*, and *plumbescens*). Although the three groups had little or no overlap, samples within each group showed a wide range of values on one or both axes. Furthermore, the variation transcended current subspecies boundaries. This was particularly evident in the interior, where ellipses of subspecies overlapped considerably. Confirming the univariate results, the smallest titmice occurred at Medford (1) and along the west slope of the Sierra Nevada (sample areas 6, 9, 13-15). The largest birds generally came from the more northern interior populations (sample areas 31, 33, 40-42, 45) and from Four Corners (50).

Males and females essentially showed the same patterns, with some exceptions. For example, the slight overlap in size among males from the central and southern Coast Ranges (sample areas 10-12, 17, 19) was not observed in females. Furthermore, females exhibited a clearer separation between samples from northern Baja California (28-29; subspecies *affabilis*) and those from southern California (18-23, 25; *transpositus*). In both sexes, birds from Joshua Tree (24; *mohavensis*) were intermediate in size and toe length between those from southern California and the interior, although they were clearly closer to the former. Likewise, individuals from Kern (16; *kernensis*) were somewhat intermediate along the first axis but also fell into group I; male *kernensis* overlapped with *sequestratus* (especially Mt. Dome [3]), whereas females grouped with *inornatus*. Males of *sequestratus* showed greater interpopulation variability than females along axis I.

Minimum spanning trees. To further assess the phenetic similarity of samples, minimum spanning trees for males (Fig. 21B) and females (Fig. 21B), calculated from Euclidean distances based on size, were superimposed onto maps showing the geographic distribution of samples. The results of this analysis generally supported the MDS plots in that titmice from the Pacific slope and interior clearly differed in size. Only a single weak branch linked these two major geographic regions (males: San Jacinto Mountains [23] to Warner [31]; females: Sierra San Pedro Mártir [29] to Southwestern Utah [51]). Pacific slope samples from southern California and northern Baja California exhibited a tight network that included the sample from Joshua Tree (24). These populations, in turn, connected to samples along the Coast Ranges in a manner consistent with the clinal variation observed in single characters. Whereas males from Santa Barbara (17) grouped closely to Ventura (19) in the MDS plot, they were linked to males from Monterey (10) in the minimum spanning tree and lacked any connection to the southern group of populations. Females showed a different picture, in that birds from Santa Barbara were most similar to those from Ventura and had no connection to the north.

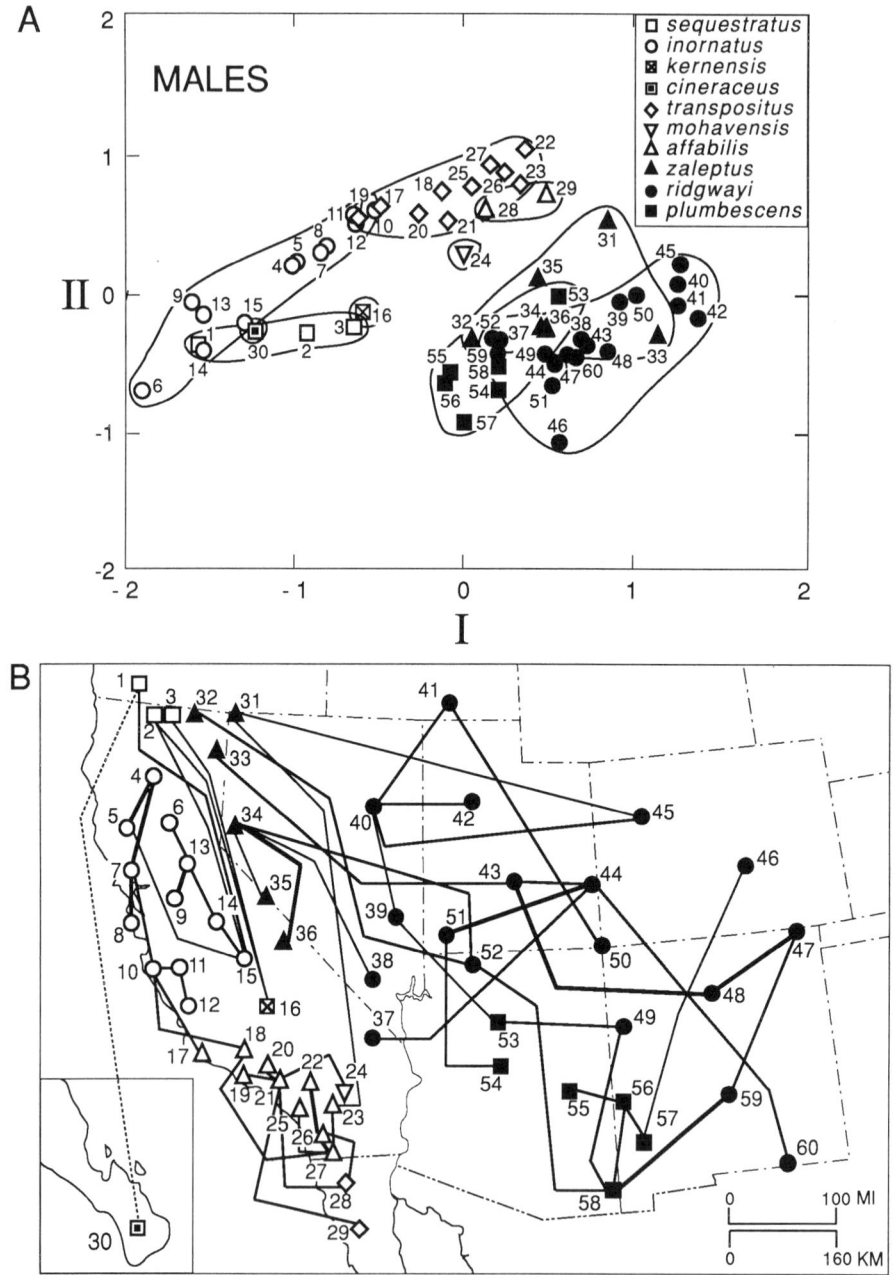

FIGURE 21. Multidimensional scaling plot (A) and minimum spanning tree (B) illustrating size variation in males among 60 samples of the *Parus inornatus* complex. Different symbols and ellipses represent subspecies recognized by the American Ornithologists' Union (1957). Samples are numbered as in Figs. 3-4 and Table 2. The final stress value for the MDS plot after 20 iterations was 0.136, indicating fair to good agreement with the original Euclidean distance matrix (Rohlf 1987). Different line weights and types in the minimum spanning tree reflect relative branch lengths and the associated strength of the connection: heavy solid line, branch length < 1; intermediate solid line, 1-1.5; light solid line, 1.5-2; light dashed line, > 2.

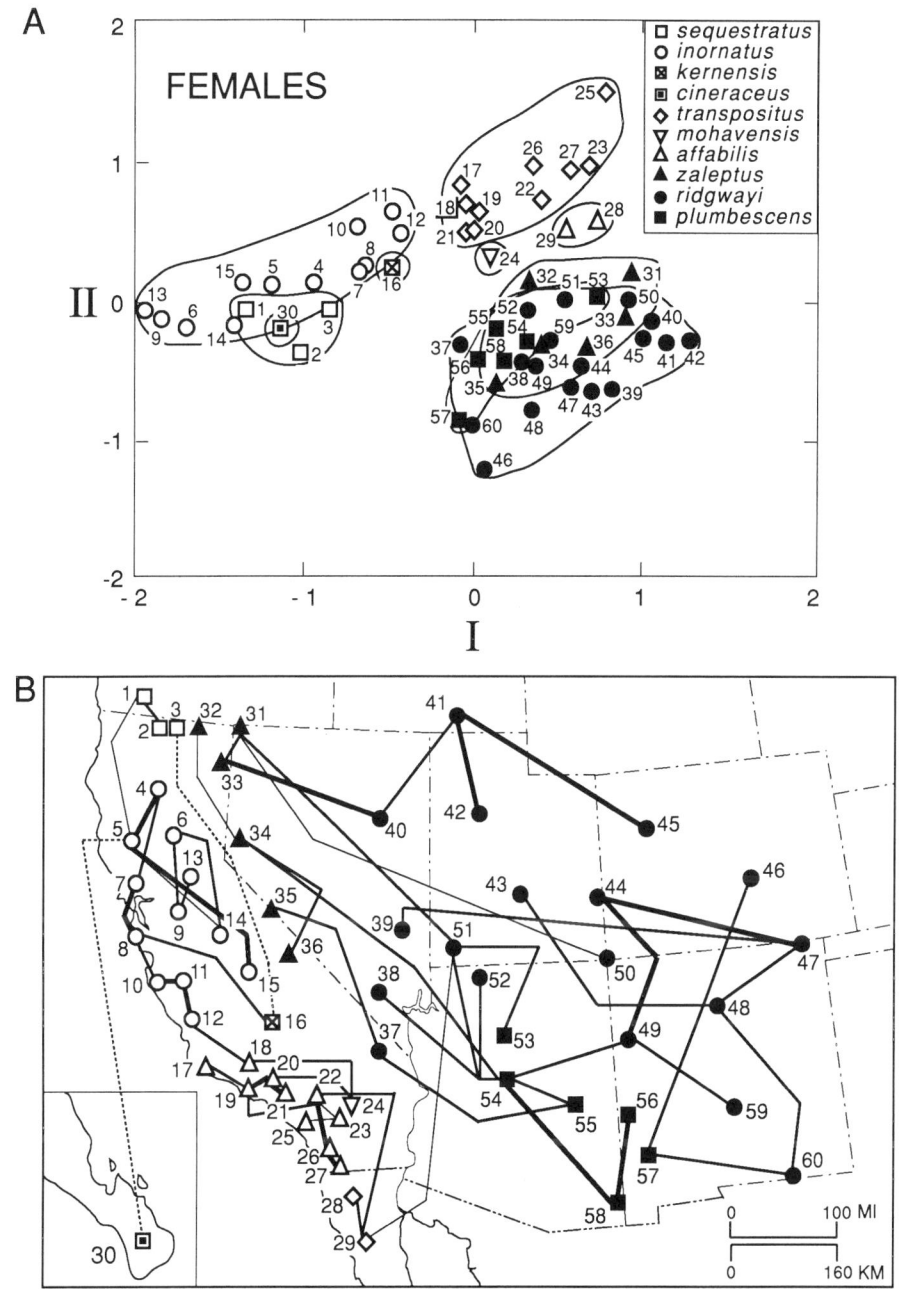

FIGURE 22. Multidimensional scaling plot (A) and minimum spanning tree (B) illustrating size variation in females among 60 samples of the *Parus inornatus* complex. Symbols for the different subspecies are the same as in Fig. 21. See legend to Fig. 21 for further detail.

Another distinct network was observed among samples from Oroville (6), Stockton (9), and the west slope of the Sierra Nevada (13-15). This chain was connected to other coastal samples by a link between the southern Sierra Nevada (15) and Clear Lake (5), a pattern not necessarily evident in the MDS plot. Males from the southern Sierra Nevada also linked weakly to those from Medford (1) and Siskiyou (2), while females from Medford showed a slight resemblance to Clear Lake. An additional connection occurred between northern California (especially Mt. Dome [3]) and Kern (16), although females from Kern were more strongly linked to those from the South San Francisco Bay Area (8) than to those from Mt. Dome. Finally, the Sierra de la Laguna (30) population was set apart from all other coastal samples, showing only a weak phenetic similarity to Medford (males) or Clear Lake (females).

In keeping with the mosaic patterns of variation exhibited by individual characters in the interior, a fairly chaotic network connected samples from the Great Basin to the Rocky Mountains and Southwest (Figs. 21B and 22B). Females showed a clearer relationship than males among populations spanning the northern part of the interior (sample areas 31, 33, 40-42, 45), where the birds tend to be larger. Furthermore, samples along the western edge of the Great Basin also were more clearly connected to each other in the network of females compared to males. Males generally showed a tighter connection among samples in southeastern Arizona and southwestern New Mexico than did females. Nonetheless, in contrast to the Pacific slope region, the general pattern was one in which geographically proximal samples often were linked to distant populations rather than or in addition to adjacent samples. For example, males from the Eastern Mojave (37) were connected to La Sal (44), while females exhibited a link both to Benton (35) and Eastern Arizona (55). Likewise, the minimum spanning tree linked males from Silver City (57) to nearby Reserve (56) and distant Pueblo (46), whereas females were allied to Pueblo and the Guadalupe Mountains (60). The disparity between males and females was greater than along the Pacific slope, where, despite minor differences, both sexes revealed the same basic patterns of size variation. While the variation among coastal populations conformed loosely to certain subspecies boundaries, the network analysis of interior populations failed to reveal a clear relationship between their phenetic similarity and subspecific classification.

Cluster analysis. The phenograms of Euclidean distances among populations unambiguously separated coastal and interior samples for males (Fig. 23) and females (Fig. 24). Within these two major clusters, several groups of populations were identified. Although males and females showed broadly similar patterns, the placement of certain populations varied between the sexes. Males from the Pacific slope grouped into three clusters comprised of (a) the smallest birds, which came from Medford (1), the east side of the Central Valley and west slope of the Sierra Nevada (6, 9, 13-15), and the Sierra de la Laguna in southern Baja California (30); (b) slightly larger birds from northern California (2-3), the North Sacramento Valley (4), the Coast and Transverse ranges from Clear Lake (5) to Newhall (20), excluding Ventura (19), and from Kern (16); and (c) medium-sized birds from southwestern California (19, 21-23, 25-27), Joshua Tree (24), and northern Baja California (28-29). Although Ventura allied with the southern group, the long branch length placed it outside of the main cluster. Likewise, samples from northern California (2-3) and Kern (16) were set off from the rest of their cluster, as was the sample from southern Baja California (30). The group comprised of the smallest birds clustered apart from all other populations.

The UPGMA also identified three interior groupings for males: (a) the largest birds, found in northern interior samples (31, 33, 40-42, 45) as well as in Southeastern Nevada (39), Four Corners (50), and the San Francisco Mountains (53); (b) moderately large birds from other samples in the western Great Basin (32, 34-36) and from the Eastern Mojave (37), Southern Nevada (38), and Kaibab (52); and (c) moderately large birds from Central Utah (43) east to Pueblo (46) and south to the Chiricahua (58) and Guadalupe (60) mountains. Males from Eastern Arizona (55) to Silver City (57) averaged slightly smaller (see MDS plot, Fig. 21A) and formed a separate subcluster.

Females from the Pacific slope were divided into two main groups that included (a) relatively small birds from Medford (1) to Kern (16) and from the Sierra de la Laguna (30), and (b) medium-sized birds from Santa Barbara (17) southward to northern Baja California (28-29). This latter group, which included Joshua Tree (24), clustered weakly with interior rather than with other coastal samples. In contrast to the pattern shown by males, females from Siskiyou (2) grouped with Medford (1), while those from Mt. Dome (3) clustered with samples from the central Coast Ranges (7-8, 10-12) and from Kern (16). The position of Santa Barbara (17) also differed between the sexes, in that females were placed with coastal samples to the south rather than with those farther north.

Within the interior, samples comprised of larger females formed a subcluster that generally agreed with the similar grouping for males (sample areas 31, 33, 40-42, 45, 39, 50, 53); females from Southwestern Utah (51) also clustered with this group instead of with La Sal (44), as exhibited by males. Another distinct cluster of females included samples from Pueblo (46), Silver City (57), and the Guadalupe Mountains (60); birds from Southern Nevada (38) were linked weakly to this group. All other interior samples formed a third cluster comprised of numerous subgroups that showed a mixed pattern in terms of geographic structuring. For example, although Reno (34) and the White-Inyo Mountains (36) allied together, Tule Lake (32) clustered with Kaibab (52), while Benton (35) grouped with the Eastern Mojave (37) sample. Similarly, females from Central Arizona (54), Reserve (56), and the Chiricahua Mountains (58) were phenetically similar to each other and also shared similarities with birds from La Sal (44), the Zuni Mountains (49), the Southwestern Plains (47), and South-central New Mexico (59); the Eastern Arizona (55) sample clustered outside of this entire group. The short branch lengths indicate weak support for some of these nodes.

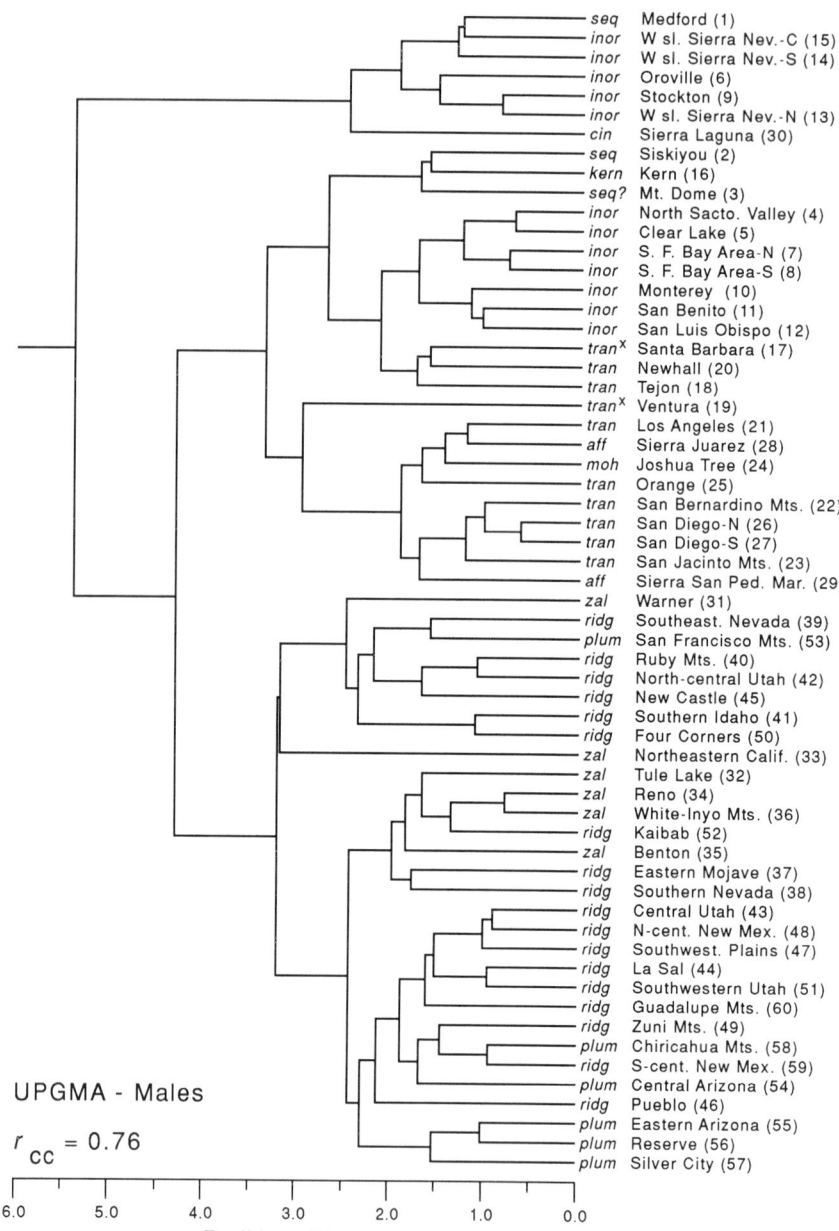

seq   Medford (1)
inor   W sl. Sierra Nev.-C (15)
inor   W sl. Sierra Nev.-S (14)
inor   Oroville (6)
inor   Stockton (9)
inor   W sl. Sierra Nev.-N (13)
cin   Sierra Laguna (30)
seq   Siskiyou (2)
kern   Kern (16)
seq?   Mt. Dome (3)
inor   North Sacto. Valley (4)
inor   Clear Lake (5)
inor   S. F. Bay Area-N (7)
inor   S. F. Bay Area-S (8)
inor   Monterey (10)
inor   San Benito (11)
inor   San Luis Obispo (12)
tran[X]   Santa Barbara (17)
tran   Newhall (20)
tran   Tejon (18)
tran[X]   Ventura (19)
tran   Los Angeles (21)
aff   Sierra Juarez (28)
moh   Joshua Tree (24)
tran   Orange (25)
tran   San Bernardino Mts. (22)
tran   San Diego-N (26)
tran   San Diego-S (27)
tran   San Jacinto Mts. (23)
aff   Sierra San Ped. Mar. (29)
zal   Warner (31)
ridg   Southeast. Nevada (39)
plum   San Francisco Mts. (53)
ridg   Ruby Mts. (40)
ridg   North-central Utah (42)
ridg   New Castle (45)
ridg   Southern Idaho (41)
ridg   Four Corners (50)
zal   Northeastern Calif. (33)
zal   Tule Lake (32)
zal   Reno (34)
zal   White-Inyo Mts. (36)
ridg   Kaibab (52)
zal   Benton (35)
ridg   Eastern Mojave (37)
ridg   Southern Nevada (38)
ridg   Central Utah (43)
ridg   N-cent. New Mex. (48)
ridg   Southwest. Plains (47)
ridg   La Sal (44)
ridg   Southwestern Utah (51)
ridg   Guadalupe Mts. (60)
ridg   Zuni Mts. (49)
plum   Chiricahua Mts. (58)
ridg   S-cent. New Mex. (59)
plum   Central Arizona (54)
ridg   Pueblo (46)
plum   Eastern Arizona (55)
plum   Reserve (56)
plum   Silver City (57)

UPGMA - Males

$r_{cc} = 0.76$

6.0   5.0   4.0   3.0   2.0   1.0   0.0

Euclidean Distance

FIGURE 23. UPGMA phenogram of Euclidean distances separating 60 samples of the *Parus inornatus* complex based on size variation in males. The cophenetic correlation coefficient (r_{cc}) indicates moderate agreement with the original distance matrix. Abbreviations for subspecific names (American Ornithologists' Union 1957) associated with each sample are indicated. Populations are numbered as in Figs. 3-4 and Table 2. The Mt. Dome sample (3), which is comprised of specimens representing new breeding localities, is assigned tentatively to the subspecies *sequestratus* (?) based on phenetic data and geographic distribution. Specimens from Santa Barbara (17) and Ventura (19), although assigned to the subspecies *transpositus*, occur in a zone of intergradation ([X]) between *transpositus* and *inornatus* (see Grinnell and Miller 1944:307-309 and Figs. 21-22, this study).

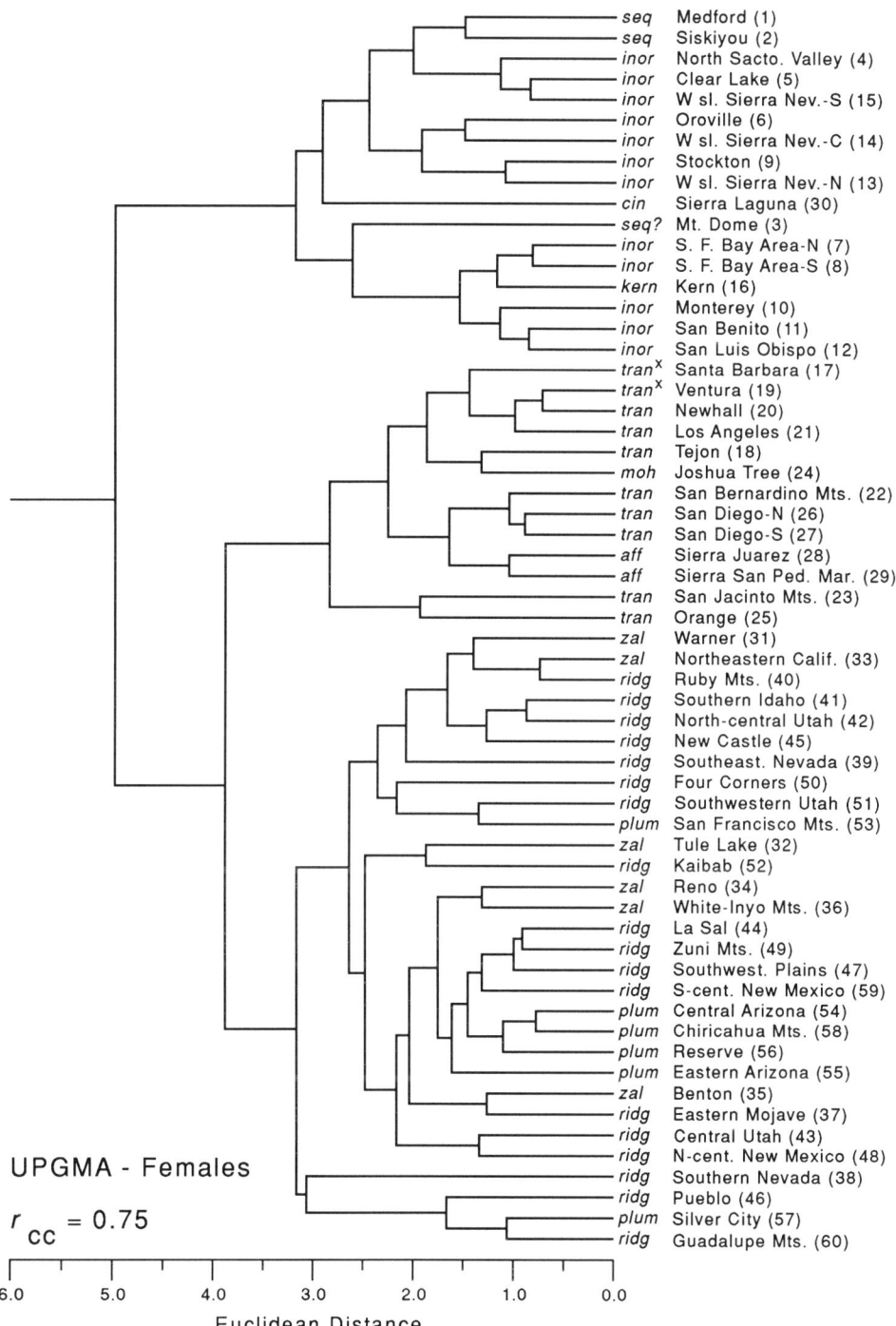

FIGURE 24. UPGMA phenogram of Euclidean distances separating 60 samples of the *Parus inornatus* complex based on size variation in females. The cophenetic correlation coefficient (r_{cc}) indicates moderate agreement with the original distance matrix. See legend to Fig. 23 for detail.

Environmental Correlates of Geographic Variation in Size

Canonical correlation analysis revealed a strong association between morphologic characters and environmental variables for both males and females (Table 10). With the exception of body weight, all morphologic characters were strongly correlated on the first canonical variate. Toe dimensions showed a different pattern from the other size characters, however, a finding that also emerged from the principal-components analysis. Although elevation loaded most heavily on the first canonical variate, several other environmental variables likewise were highly correlated with morphology (longitude, average minimum December temperature, December mean temperature, average minimum January temperature, January mean temperature, seasonal temperature difference, mean August precipitation, mean December precipitation, and mean January precipitation). This analysis clearly reflects an increase in overall size of titmice toward the interior (i.e., lower longitude), where populations occupy habitats at higher elevations and where winter temperatures and seasonality are more extreme. Titmice in those areas also experience drier winters and increased summer precipitation compared to birds along the Pacific slope. The higher correlation of winter versus summer temperatures on morphology was especially interesting in light of the fact that growth occurs primarily during the spring and summer months. In a similar analysis for Fox Sparrows, Zink (1986) found a moderate correlation between May and June temperatures and morphology, and suggested that temperatures during these months might influence growth and thus adult morphology. Unlike Fox Sparrows, however, titmice are non-migratory, and different factors might be expected to influence morphologic variation in resident versus non-resident bird species.

GEOGRAPHIC VARIATION IN PLUMAGE COLORATION

Univariate Patterns of Color Variation

Geographic variation in color was examined for three parameters (dominant wavelength, brightness, and purity) measured on each of three body regions (dorsum, breast, and belly). Because color on different body regions may show similar patterns, I first used linear regression analysis to correlate sample means of body color for six pairwise comparisons (Fig. 25). Inter-population variability in breast and belly color was highly correlated for all three parameters. In contrast, dorsal coloration appeared to vary independently of breast and belly color for either dominant wavelength or brightness. Purity, on the other hand, showed a moderate geographic correlation between the dorsal and ventral plumage. In light of these findings, results are presented only for the dorsum and breast. Patterns of variation in belly color presumably parallel those for the breast, given the fairly strong correlation between them. Data on color were not available for all geographic samples included in the analysis of size variation (see Table 2 and "Materials and Methods").

TABLE 10. Canonical correlation analysis on 26 environmental variables and 9 morphologic characters for males and females. Values indicate the correlation between each character in the two data sets and the first canonical variate. The canonical correlation between the morphologic and environmental data sets on the first canonical variate is 0.980 for males and 0.977 for females.

Variable	Males	Females	Variable	Males	Females
Environmental					
Elevation	0.946	-0.925	Dec. Temp.-Min.	-0.849	0.861
Latitude	0.029	-0.068	Dec. Temp.-Mean	-0.769	0.784
Longitude	-0.828	0.784	Jan. Temp.-Max.	-0.655	0.672
April Temp.-Max	-0.345	0.375	Jan. Temp.-Min.	-0.848	0.859
April Temp.-Min	-0.675	0.694	Jan. Temp.-Mean	-0.776	0.790
April Temp.-Mean	-0.560	0.586	Seasonal Temp. Diff.[a]	0.794	-0.783
May Temp.-Max.	-0.107	0.137	April Precip.	-0.645	0.645
May Temp.-Min.	-0.499	0.528	May Precip.	0.370	-0.354
May Temp.-Mean	-0.342	0.374	June Precip.	0.627	-0.614
June Temp.-Max.	0.151	-0.113	August Precip.	0.728	-0.689
June Temp.-Min.	-0.266	0.309	Dec. Precip.	-0.793	0.791
June Temp.-Mean	-0.062	0.104	Jan. Precip.	-0.843	0.828
Dec. Temp.-Max.	-0.629	0.646	Annual Precip.	-0.605	0.615
Morphological					
Wing Length	0.899	-0.889	Bill Depth	0.568	-0.425
Tail Length	0.708	-0.617	Bill Width	0.859	-0.802
Tarsus Length	0.526	-0.381	Bill Length	0.840	-0.848
Length of Middle Toe	-0.541	0.497	Body Wt (cube root)	0.130	-0.074
Hallux Length	-0.588	0.573			

[a] Calculated as the difference between June mean temperature and December mean temperature.

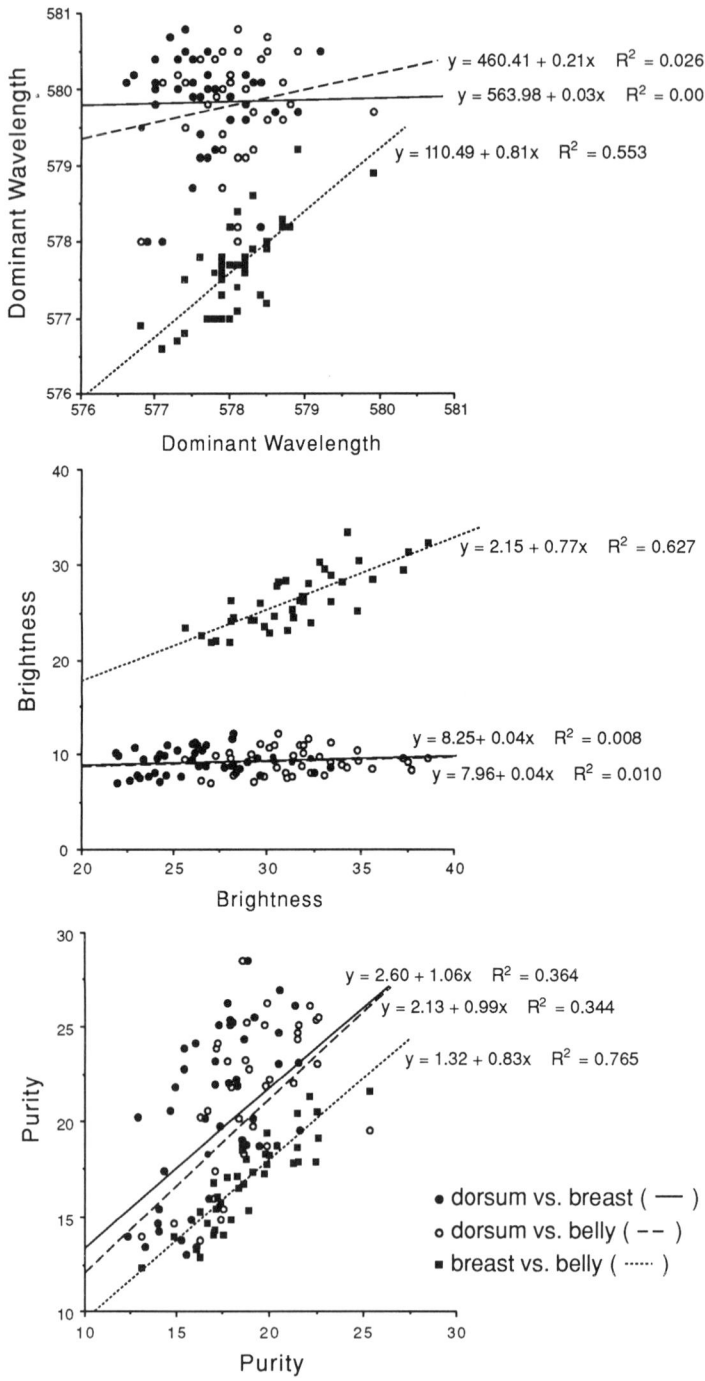

FIGURE 25. Bivariate regression plots showing variation in plumage color for three regions of the body (dorsum, breast, belly) as measured by dominant wavelength, brightness, and purity. Symbols and lines illustrate different pairwise comparisons. Each point represents the mean for a single sample area.

Dorsal color (Figs. 26-28). Dominant wavelength, brightness, and purity of the dorsum all varied geographically between coastal and interior samples as well as within each region. An abrupt change in dominant wavelength (Fig. 26) occurred across the desert from Joshua Tree (24) to the Eastern Mojave (37). With a few exceptions (sample areas 45, 51, 53), dominant wavelength generally was lower throughout the interior compared to the Pacific slope. Interior samples showed substantial local variability in dominant wavelength, however, resulting in a mosaic pattern. Titmice from North-central Utah (42) resembled those from the Eastern Mojave in having extremely low values for dominant wavelength.

Clearer trends were exhibited by variation in brightness and purity of the dorsum. Means for brightness (Fig. 27) were clearly higher in the interior than along the Pacific slope, particularly in the more northern populations (sample areas 33, 41-43). Birds from North-central New Mexico (48) also were characterized by relatively high brightness of the dorsum, as were individuals from the Eastern Mojave (37). Within the Southwest, brightness generally declined from northwest to southeast (51-60). A significant reduction in brightness occurred between the Eastern Mojave and Joshua Tree (24), which in turn had higher values compared to adjacent coastal samples (22-23). In general, titmice from southern California and northern Baja California (19-29) had darker plumage (i.e., low brightness) relative to birds from populations farther north. A major break occurred in the vicinity of Ventura (19) and Newhall (20), with birds from Santa Barbara (17) and Tejon (18) having medium values for brightness similar to those from San Luis Obispo (12) and Kern (16). From there, brightness declined northward, although there was some local variability. Brightness was higher in Siskiyou (2) than in Medford (1) or the North Sacramento Valley (4), but all of these samples had much lower values compared to Northeastern California (33). Titmice from the Sierra de la Laguna (30) were more similar to birds from the central Coast Range (e.g., sample areas 10-11) or Medford than to those from northern Baja California.

Purity of the dorsum (Fig. 28) showed a similar trend to dominant wavelength in that, unlike brightness, values were lower in the interior than along the Pacific slope. The lowest values occurred in the Eastern Mojave (37) and from southern Idaho to southwestern Utah and the Kaibab Plateau (41-42, 51-52). Titmice from the Southwestern Plains (47) and North-central New Mexico (48) also were characterized by low purity compared to adjacent samples. Higher values of purity were found in Central Utah (43), Colorado (45-46), and the southern samples of the interior (53-60). In concordance with dominant wavelength and brightness, purity changed abruptly between the Eastern Mojave (37) and Joshua Tree (24). Birds from Joshua Tree again were intermediate compared to those from the San Bernardino Mountains (22), San Jacinto Mountains (23), and other coastal populations. Unlike brightness, however, purity did not differ noticeably between northern and southern samples within the Pacific slope region. Birds from this region had relatively high values compared to individuals from eastern California (33, 36).

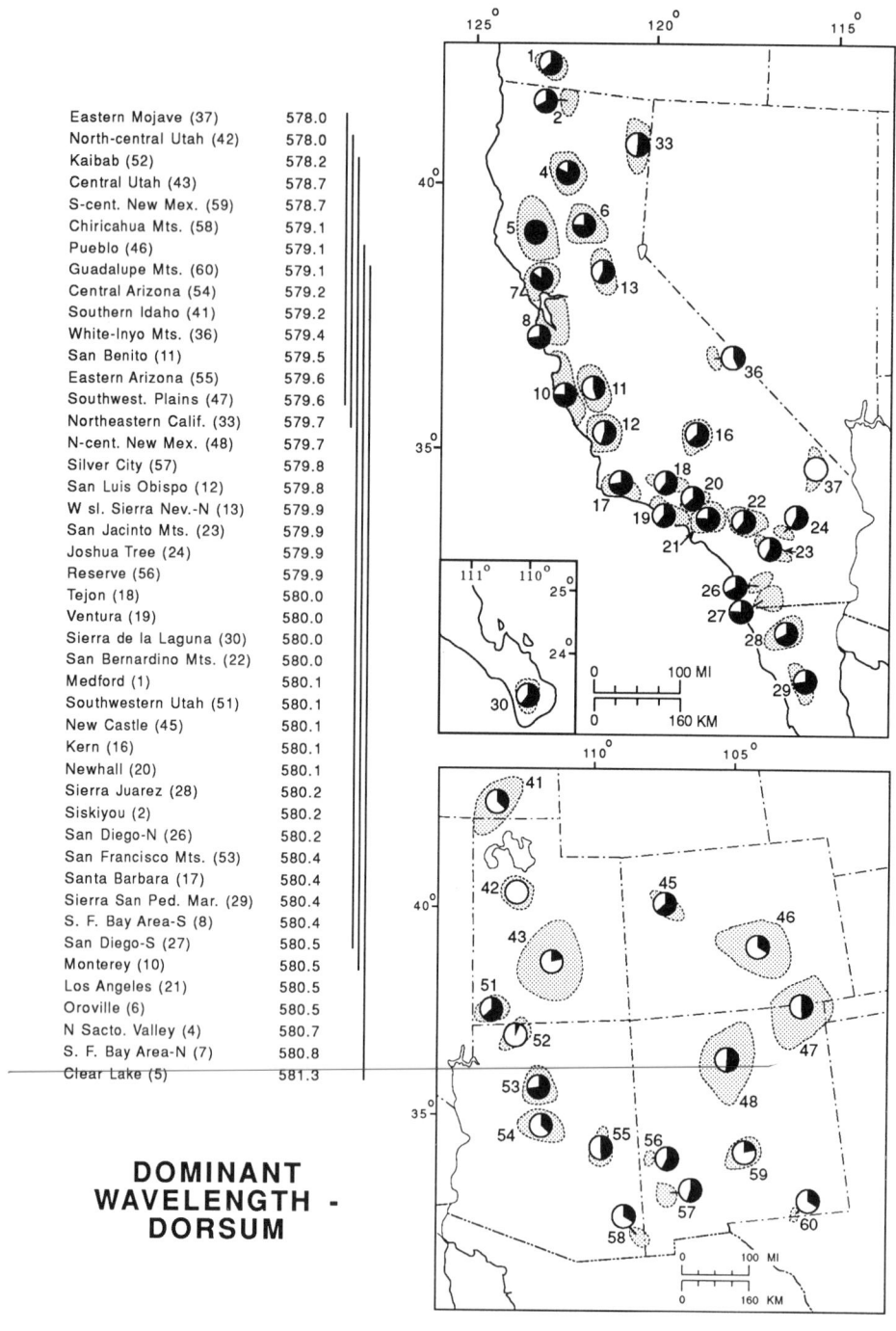

Eastern Mojave (37)	578.0
North-central Utah (42)	578.0
Kaibab (52)	578.2
Central Utah (43)	578.7
S-cent. New Mex. (59)	578.7
Chiricahua Mts. (58)	579.1
Pueblo (46)	579.1
Guadalupe Mts. (60)	579.1
Central Arizona (54)	579.2
Southern Idaho (41)	579.2
White-Inyo Mts. (36)	579.4
San Benito (11)	579.5
Eastern Arizona (55)	579.6
Southwest. Plains (47)	579.6
Northeastern Calif. (33)	579.7
N-cent. New Mex. (48)	579.7
Silver City (57)	579.8
San Luis Obispo (12)	579.8
W sl. Sierra Nev.-N (13)	579.9
San Jacinto Mts. (23)	579.9
Joshua Tree (24)	579.9
Reserve (56)	579.9
Tejon (18)	580.0
Ventura (19)	580.0
Sierra de la Laguna (30)	580.0
San Bernardino Mts. (22)	580.0
Medford (1)	580.1
Southwestern Utah (51)	580.1
New Castle (45)	580.1
Kern (16)	580.1
Newhall (20)	580.1
Sierra Juarez (28)	580.2
Siskiyou (2)	580.2
San Diego-N (26)	580.2
San Francisco Mts. (53)	580.4
Santa Barbara (17)	580.4
Sierra San Ped. Mar. (29)	580.4
S. F. Bay Area-S (8)	580.4
San Diego-S (27)	580.5
Monterey (10)	580.5
Los Angeles (21)	580.5
Oroville (6)	580.5
N Sacto. Valley (4)	580.7
S. F. Bay Area-N (7)	580.8
Clear Lake (5)	581.3

DOMINANT WAVELENGTH - DORSUM

FIGURE 26. SS-STP applied to mean dominant wavelength of the dorsum for 45 samples of the *Parus inornatus* complex (males and females combined). The analysis and presentation of geographic variation in this character is the same as for wing length (see legend to Fig. 13).

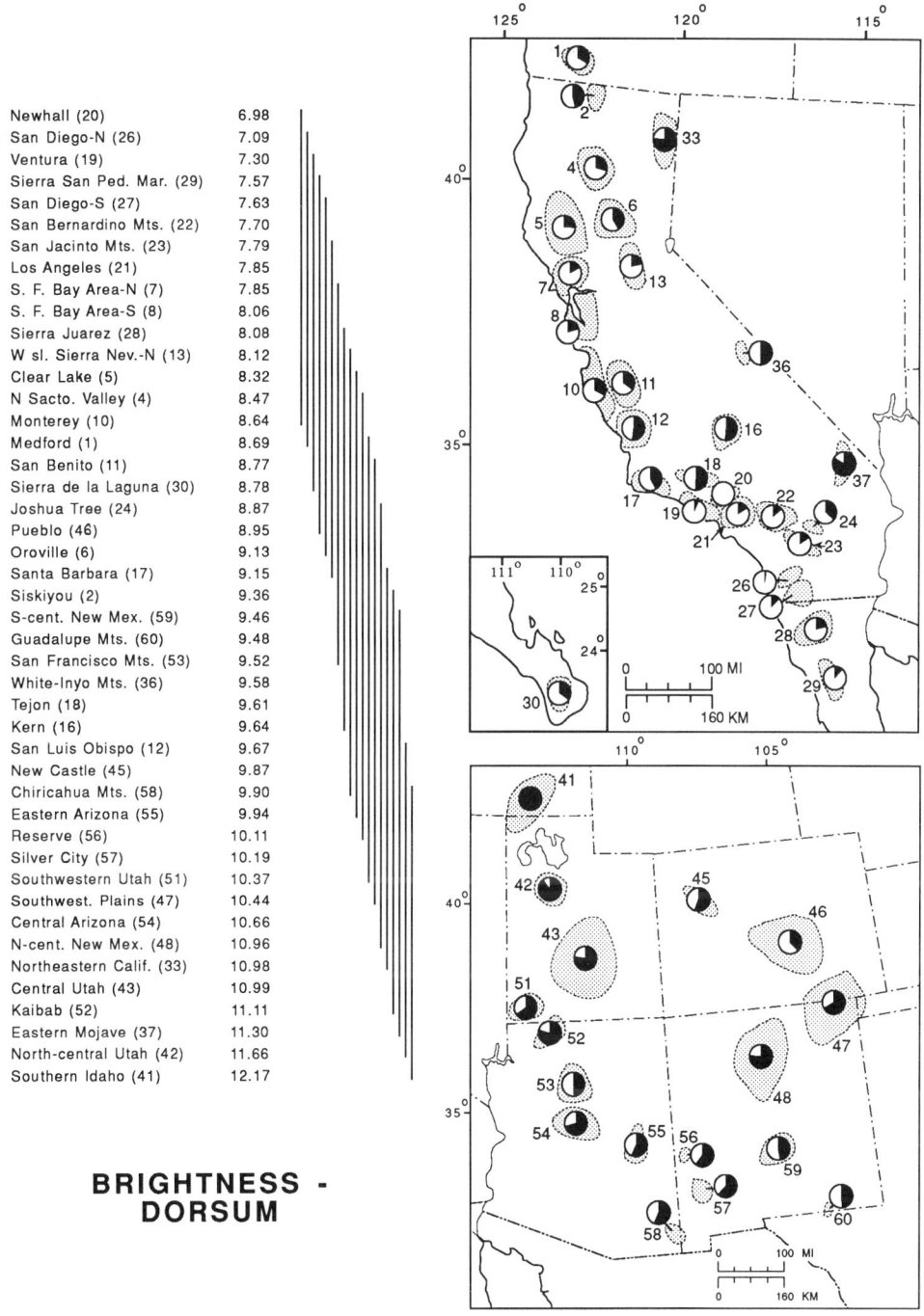

Newhall (20)	6.98
San Diego-N (26)	7.09
Ventura (19)	7.30
Sierra San Ped. Mar. (29)	7.57
San Diego-S (27)	7.63
San Bernardino Mts. (22)	7.70
San Jacinto Mts. (23)	7.79
Los Angeles (21)	7.85
S. F. Bay Area-N (7)	7.85
S. F. Bay Area-S (8)	8.06
Sierra Juarez (28)	8.08
W sl. Sierra Nev.-N (13)	8.12
Clear Lake (5)	8.32
N Sacto. Valley (4)	8.47
Monterey (10)	8.64
Medford (1)	8.69
San Benito (11)	8.77
Sierra de la Laguna (30)	8.78
Joshua Tree (24)	8.87
Pueblo (46)	8.95
Oroville (6)	9.13
Santa Barbara (17)	9.15
Siskiyou (2)	9.36
S-cent. New Mex. (59)	9.46
Guadalupe Mts. (60)	9.48
San Francisco Mts. (53)	9.52
White-Inyo Mts. (36)	9.58
Tejon (18)	9.61
Kern (16)	9.64
San Luis Obispo (12)	9.67
New Castle (45)	9.87
Chiricahua Mts. (58)	9.90
Eastern Arizona (55)	9.94
Reserve (56)	10.11
Silver City (57)	10.19
Southwestern Utah (51)	10.37
Southwest. Plains (47)	10.44
Central Arizona (54)	10.66
N-cent. New Mex. (48)	10.96
Northeastern Calif. (33)	10.98
Central Utah (43)	10.99
Kaibab (52)	11.11
Eastern Mojave (37)	11.30
North-central Utah (42)	11.66
Southern Idaho (41)	12.17

BRIGHTNESS - DORSUM

FIGURE 27.   SS-STP applied to mean brightness of the dorsum for 45 samples of the *Parus inornatus* complex (males and females combined).   The analysis and presentation of geographic variation in this character is the same as for wing length (see legend to Fig. 13).

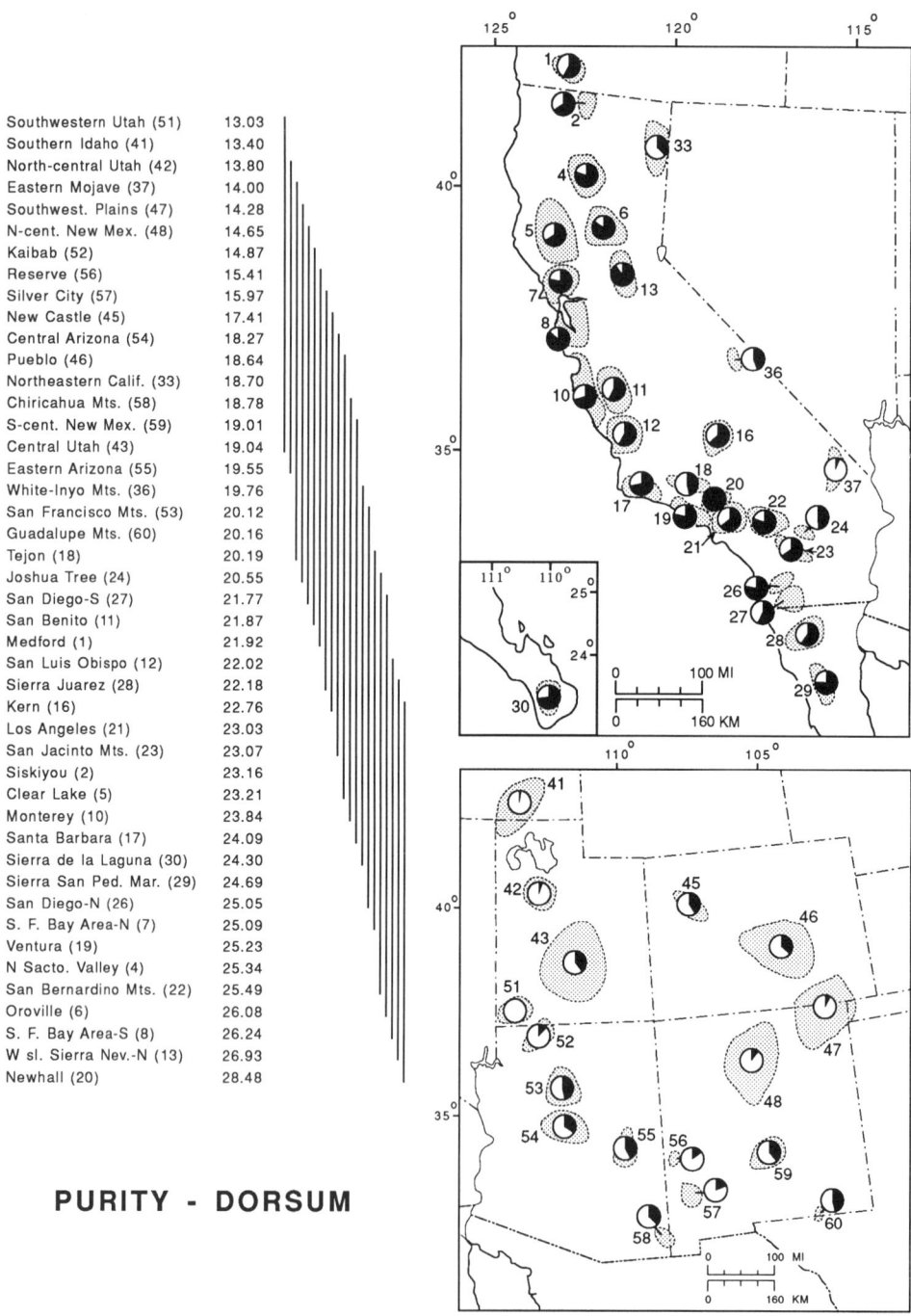

Southwestern Utah (51)	13.03
Southern Idaho (41)	13.40
North-central Utah (42)	13.80
Eastern Mojave (37)	14.00
Southwest. Plains (47)	14.28
N-cent. New Mex. (48)	14.65
Kaibab (52)	14.87
Reserve (56)	15.41
Silver City (57)	15.97
New Castle (45)	17.41
Central Arizona (54)	18.27
Pueblo (46)	18.64
Northeastern Calif. (33)	18.70
Chiricahua Mts. (58)	18.78
S-cent. New Mex. (59)	19.01
Central Utah (43)	19.04
Eastern Arizona (55)	19.55
White-Inyo Mts. (36)	19.76
San Francisco Mts. (53)	20.12
Guadalupe Mts. (60)	20.16
Tejon (18)	20.19
Joshua Tree (24)	20.55
San Diego-S (27)	21.77
San Benito (11)	21.87
Medford (1)	21.92
San Luis Obispo (12)	22.02
Sierra Juarez (28)	22.18
Kern (16)	22.76
Los Angeles (21)	23.03
San Jacinto Mts. (23)	23.07
Siskiyou (2)	23.16
Clear Lake (5)	23.21
Monterey (10)	23.84
Santa Barbara (17)	24.09
Sierra de la Laguna (30)	24.30
Sierra San Ped. Mar. (29)	24.69
San Diego-N (26)	25.05
S. F. Bay Area-N (7)	25.09
Ventura (19)	25.23
N Sacto. Valley (4)	25.34
San Bernardino Mts. (22)	25.49
Oroville (6)	26.08
S. F. Bay Area-S (8)	26.24
W sl. Sierra Nev.-N (13)	26.93
Newhall (20)	28.48

PURITY - DORSUM

FIGURE 28. SS-STP applied to mean purity of the dorsum for 45 samples of the *Parus inornatus* complex (males and females combined). The analysis and presentation of geographic variation in this character is the same as for wing length (see legend to Fig. 13).

Breast color (Figs. 29-31). Breast color showed a greater mosaic of patterns compared to dorsal color in all three parameters. Dominant wavelength (Fig. 29) was lowest in samples from Medford (1) and Siskiyou (2), varying significantly from the Northeastern California (33) population. Birds from the North Sacramento Valley (4) also had relatively low values for dominant wavelength. Another break occurred between the Eastern Mojave (37) and Joshua Tree (24), which in turn had higher values compared to the San Bernardino Mountains (22) but lower values relative to the San Jacinto range (23). Although birds from southwestern California, Baja California, and the interior had mostly medium to high values, local variability among populations confounds the general pattern.

As with dorsal plumage, brightness of the breast (Fig. 30) was notably lower in southern California and northern Baja California (19-23, 26-29) compared to samples farther north (1-18). Importantly, the sharp change in breast color coincided geographically with variation in brightness of the dorsum. Birds from the Sierra de la Laguna (30) had higher values than titmice from other samples in Baja California, whereas those from Joshua Tree (24) had medium values that exceeded those found in nearby coastal samples or in the Eastern Mojave (37). Toward the north end of the range, brightness was clearly higher in birds from Siskiyou (2) compared to those from Medford (1) and especially Northeastern California (33). Values for Northeastern California were higher than for the White-Inyo Mountains (36), however, where mean brightness of the breast was similar to that found in southern California, northern Baja California, and the Southwest. Sample means for the northern Great Basin (41-42) were higher than for elsewhere in the interior, but less than those from the central to northern Pacific slope.

Although breast purity (Fig. 31) varied more chaotically than brightness, several clear differences emerged. The most notable of these is the sharp contrast between the Eastern Mojave (37), which had the lowest mean, and the San Jacinto Mountains (23), where breast purity reached the other extreme. Birds from Newhall (20), Los Angeles (21), and the San Bernardino Mountains (22) also had relatively high values, while purity at Joshua Tree (24) averaged much lower. Breast purity increased from the Eastern Mojave to Northeastern California (33), which had higher values compared to Medford (1), Siskiyou (2), and the North Sacramento Valley (4). Other relatively high values for purity were obtained along the northwestern slope of the Sierra Nevada (6, 13) and in Baja California (28-30). Within the interior, purity varied from a low in Southern Idaho (41) to a high in Eastern Arizona (55). With some exceptions, the more southerly interior samples generally had higher means for breast purity compared to those toward the north.

Medford (1)	576.6
Siskiyou (2)	576.7
San Benito (11)	576.8
North-central Utah (42)	576.9
Tejon (18)	577.0
S. F. Bay Area-S (8)	577.0
San Luis Obispo (12)	577.0
Kern (16)	577.0
Eastern Mojave (37)	577.1
N Sacto. Valley (4)	577.2
Santa Barbara (17)	577.3
San Bernardino Mts. (22)	577.3
S. F. Bay Area-N (7)	577.4
Monterey (10)	577.4
Central Utah (43)	577.5
Newhall (20)	577.5
San Francisco Mts. (53)	577.5
W sl. Sierra Nev.-N (13)	577.5
Joshua Tree (24)	577.6
Reserve (56)	577.6
Chiricahua Mts. (58)	577.6
White-Inyo Mts. (36)	577.6
Los Angeles (21)	577.7
San Diego-N (26)	577.7
Sierra de la Laguna (30)	577.7
Guadalupe Mts. (60)	577.7
Southern Idaho (41)	577.8
Sierra San Ped. Mar. (29)	577.8
Ventura (19)	577.8
Oroville (6)	577.9
Central Arizona (54)	577.9
San Jacinto Mts. (23)	578.0
Eastern Arizona (55)	578.0
Sierra Juarez (28)	578.2
Southwest. Plains (47)	578.2
Silver City (57)	578.2
New Castle (45)	578.3
Southwestern Utah (51)	578.4
Kaibab (52)	578.4
Northeastern Calif. (33)	578.6
N-cent. New Mex. (48)	578.9
San Diego-S (27)	579.2

DOMINANT
WAVELENGTH -
BREAST

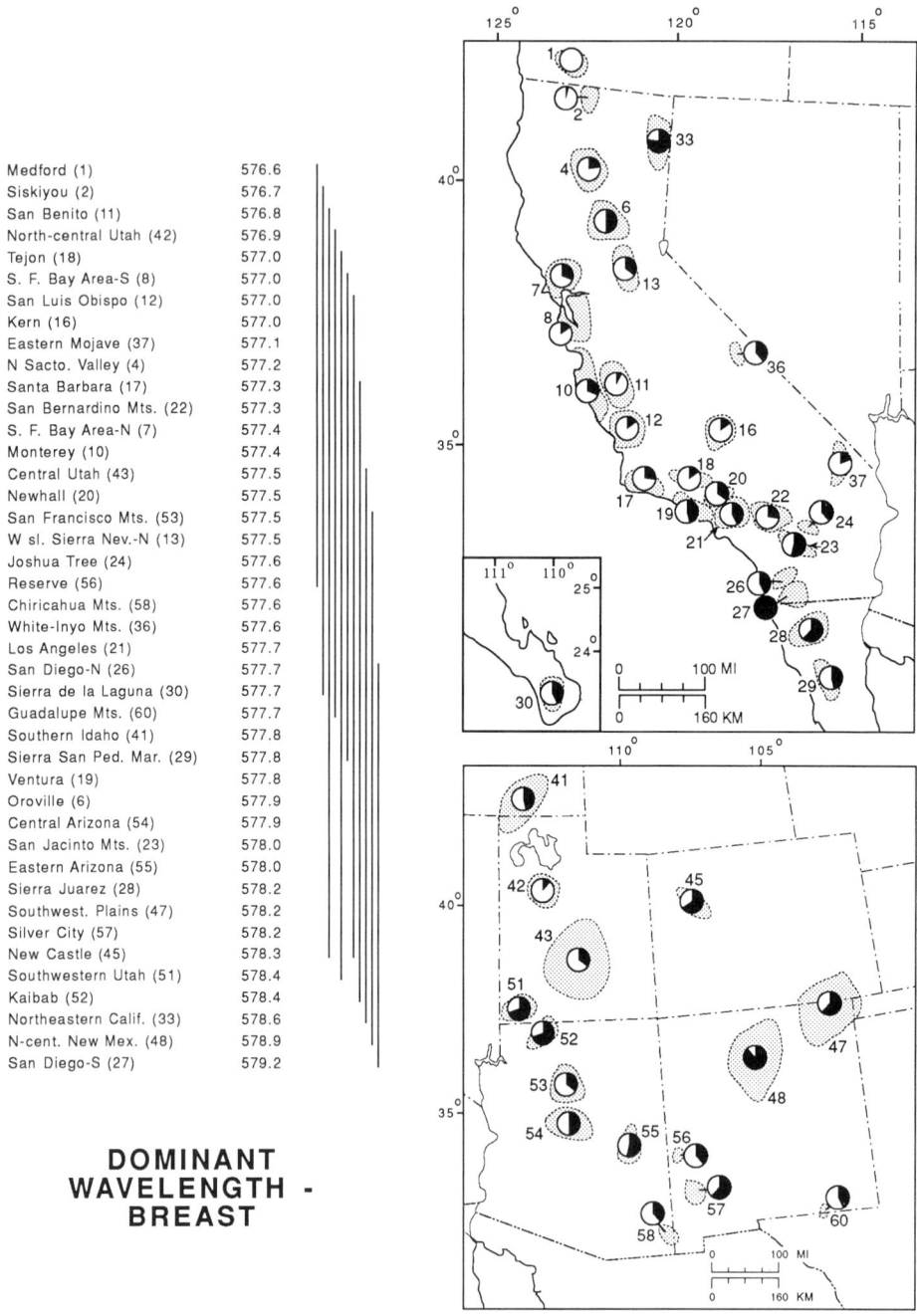

FIGURE 29. SS-STP applied to mean dominant wavelength of the breast for 42 samples of the *Parus inornatus* complex (males and females combined). The analysis and presentation of geographic variation in this character is the same as for wing length (see legend to Fig. 13). The irregular pattern of lines linking subsets 3 (from 576.8 to 578.3) and 5 (from 577.0 to 578.4) is attributed to small and uneven sample sizes (Sokal and Rohlf 1981:260; also Sokal, pers. comm.). Re-analysis of the data by increasing sample sizes to a minimum of 10, and by making all sample sizes even, resulted in a smooth pattern of lines such as that observed in other figures.

Reserve (56)	21.87
Newhall (20)	21.96
Chiricahua Mts. (58)	21.99
Ventura (19)	22.63
Central Arizona (54)	22.91
San Jacinto Mts. (23)	23.00
Sierra San Ped. Mar. (29)	23.19
Guadalupe Mts. (60)	23.36
San Bernardino Mts. (22)	23.63
Sierra Juarez (28)	23.98
White-Inyo Mts. (36)	24.12
Eastern Arizona (55)	24.22
San Diego-N (26)	24.23
New Castle (45)	24.53
Los Angeles (21)	24.58
Central Utah (43)	24.65
Southwest. Plains (47)	25.19
San Diego-S (27)	25.37
Kaibab (52)	25.96
San Francisco Mts. (53)	25.97
Eastern Mojave (37)	26.12
Silver City (57)	26.13
N-cent. New Mex. (48)	26.27
San Benito (11)	26.30
Southwestern Utah (51)	26.50
Northeastern Calif. (33)	26.74
Sierra de la Laguna (30)	26.74
Medford (1)	27.75
North-central Utah (42)	28.12
Southern Idaho (41)	28.20
Joshua Tree (24)	28.22
S. F. Bay Area-S (8)	28.38
N Sacto. Valley (4)	28.52
Oroville (6)	28.93
Tejon (18)	29.50
S. F. Bay Area-N (7)	29.60
San Luis Obispo (12)	30.29
Siskiyou (2)	30.38
Santa Barbara (17)	31.32
Kern (16)	32.35
W sl. Sierra Nev.-N (13)	32.54
Monterey (10)	33.43

BRIGHTNESS - BREAST

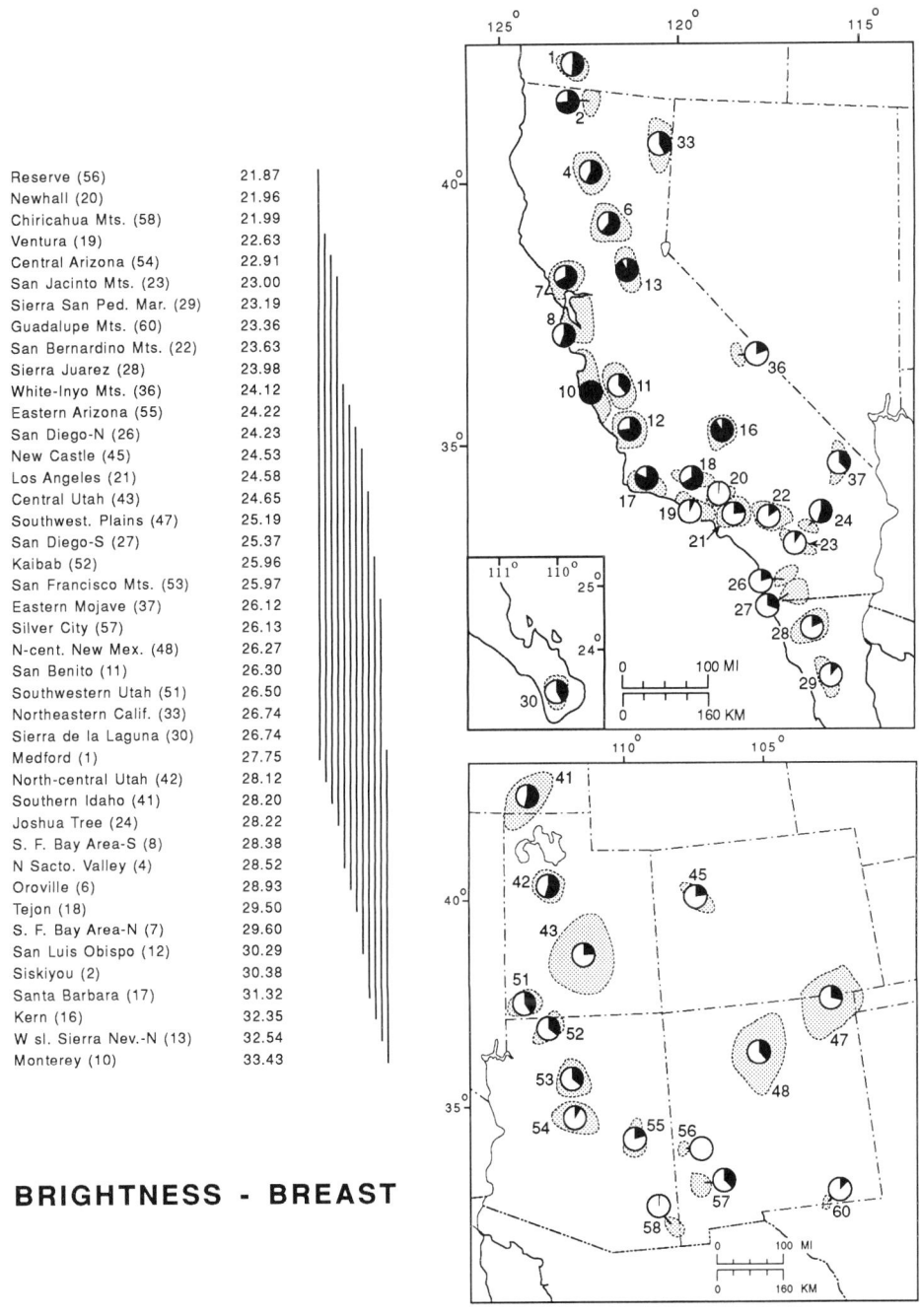

FIGURE 30. SS-STP applied to mean brightness of the breast for 42 samples of the *Parus inornatus* complex (males and females combined). The analysis and presentation of geographic variation in this character is the same as for wing length (see legend to Fig. 13).

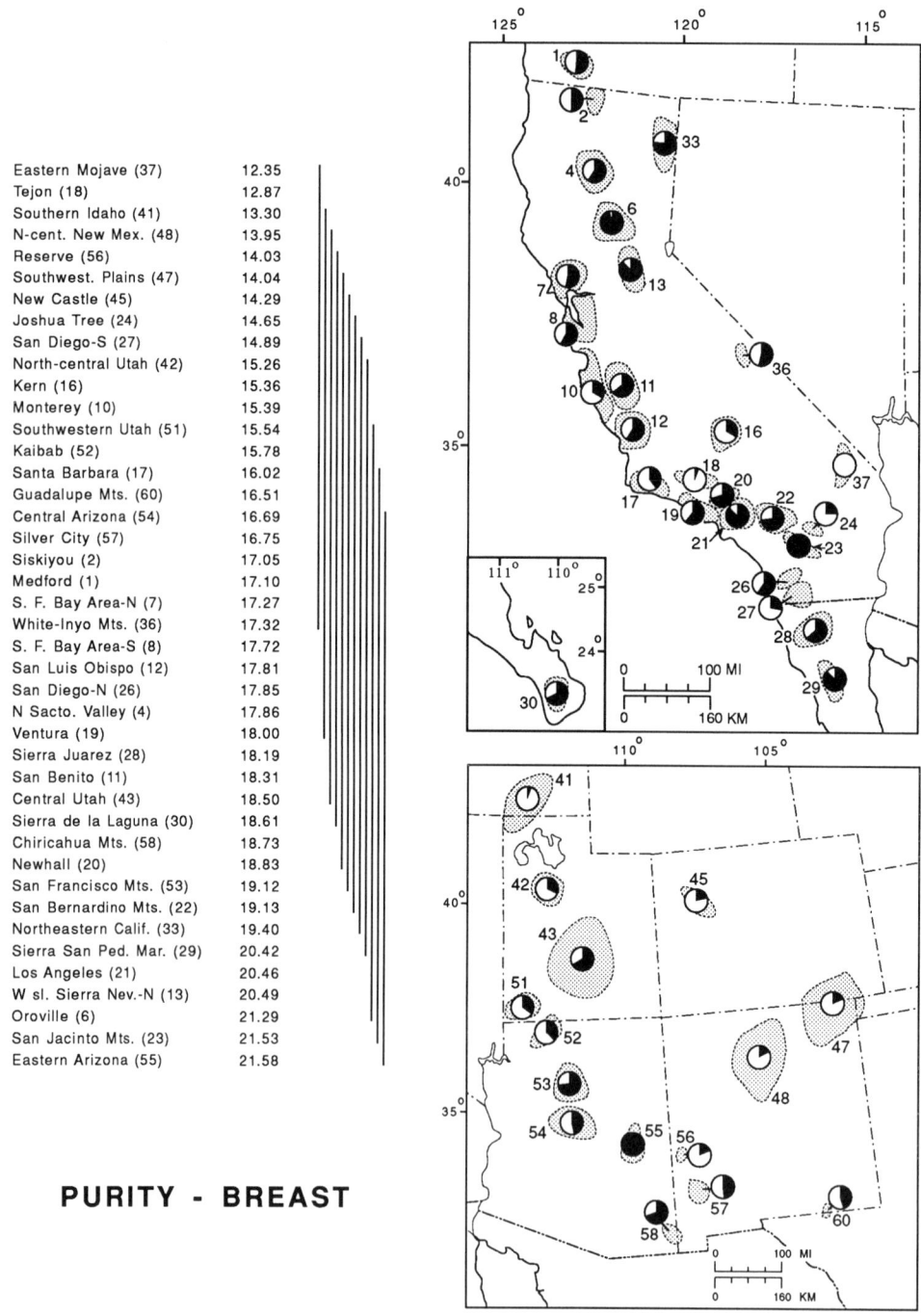

Eastern Mojave (37)   12.35
Tejon (18)   12.87
Southern Idaho (41)   13.30
N-cent. New Mex. (48)   13.95
Reserve (56)   14.03
Southwest. Plains (47)   14.04
New Castle (45)   14.29
Joshua Tree (24)   14.65
San Diego-S (27)   14.89
North-central Utah (42)   15.26
Kern (16)   15.36
Monterey (10)   15.39
Southwestern Utah (51)   15.54
Kaibab (52)   15.78
Santa Barbara (17)   16.02
Guadalupe Mts. (60)   16.51
Central Arizona (54)   16.69
Silver City (57)   16.75
Siskiyou (2)   17.05
Medford (1)   17.10
S. F. Bay Area-N (7)   17.27
White-Inyo Mts. (36)   17.32
S. F. Bay Area-S (8)   17.72
San Luis Obispo (12)   17.81
San Diego-N (26)   17.85
N Sacto. Valley (4)   17.86
Ventura (19)   18.00
Sierra Juarez (28)   18.19
San Benito (11)   18.31
Central Utah (43)   18.50
Sierra de la Laguna (30)   18.61
Chiricahua Mts. (58)   18.73
Newhall (20)   18.83
San Francisco Mts. (53)   19.12
San Bernardino Mts. (22)   19.13
Northeastern Calif. (33)   19.40
Sierra San Ped. Mar. (29)   20.42
Los Angeles (21)   20.46
W sl. Sierra Nev.-N (13)   20.49
Oroville (6)   21.29
San Jacinto Mts. (23)   21.53
Eastern Arizona (55)   21.58

PURITY - BREAST

FIGURE 31. SS-STP applied to mean purity of the breast for 42 samples of the *Parus inornatus* complex (males and females combined). The analysis and presentation of geographic variation in this character is the same as for wing length (see legend to Fig. 13).

Multivariate Patterns of Color Variation

Ordination analysis. In a principal-components analysis of the 9 color variables, 82.9% of the variance was explained by the first three axes (Table 11). Factor loadings on PC 1 reflected differences in dorsal coloration (dominant wavelength, purity, and brightness) and in purity of the breast and belly. Brightness of ventral plumage loaded most heavily on PC 2, while measures of dominant wavelength comprised the third axis. Brightness and purity were negatively correlated with each other in all body regions.

To better understand the patterns of variation among samples, data on the color of back and breast plumage were analyzed separately using principal-components analysis (Table 12) and multidimensional scaling (Fig. 32). Belly measurements were omitted from these analyses because of the strong correlation between breast and belly color (see Fig. 25). The MDS plots yielded different groupings of samples based on dorsal (Fig. 32A) versus breast (Fig. 32B) coloration. Pacific slope and interior samples showed clear separation in dorsal color along axis I, which had high positive loadings for dominant wavelength and purity and a high negative loading for brightness (Table 12). Most of the variance (81.5%) was explained by this first axis, whereas populations showed little separation along axis II (13.9% of variance). Birds from the San Francisco Mountains (53) had slightly darker back plumage that tended toward the color found in coastal populations. Without this sample, complete separation between the two groups based on dorsal coloration would have been achieved. Samples assigned to current subspecies overlapped considerably in each group.

TABLE 11. Character loadings on the first three principal components of variation in 3 measures of color of the dorsum, breast, and belly in the *Parus inornatus* complex[a].

	PC 1	PC 2	PC 3
Dominant Wavelength-Dorsum	0.662	0.322	0.533
Dominant Wavelength-Breast	-0.227	-0.567	0.723
Dominant Wavelength-Belly	-0.180	-0.160	0.920
Purity-Dorsum	0.917	0.238	0.002
Purity-Breast	0.804	-0.330	-0.091
Purity-Belly	0.813	-0.268	-0.001
Brightness-Dorsum	-0.866	-0.021	-0.106
Brightness-Breast	-0.028	0.897	0.052
Brightness-Belly	-0.115	0.853	0.359
Variance Explained	38.16%	24.66%	20.04%

[a] Males, females, and unsexed birds were combined for the analysis. Loadings are based on a variance-covariance matrix of mean values (standardized by variance) for 45 samples.

TABLE 12. Character loadings on the first three principal components of color variation in the *Parus inornatus* complex[a] based on separate analyses of the dorsum and breast.

	PC 1	PC 2	PC 3
Dorsum			
Dominant Wavelength	0.839	-0.544	0.018
Purity	0.928	0.270	0.256
Brightness	-0.938	-0.219	0.270
Variance Explained	81.51%	13.88%	4.61%
Breast			
Dominant Wavelength	0.756	-0.459	0.467
Purity	0.353	0.899	0.258
Brightness	-0.853	-0.034	0.520
Variance Explained	47.50%	34.01%	18.49%

[a] Males, females, and unsexed birds were combined for the analysis. Loadings are based on a variance-covariance matrix of sample means standardized by variance. Data on the color of the dorsum were taken for 45 samples; breast measurements were available for 42 samples (samples with missing data included Clear Lake [5], Pueblo [46], and south-central New Mexico [59]).

Breast color showed greater overlap than dorsal color among samples and subspecies from different geographic regions (Fig. 32B). However, birds from Medford (1) and northern to central California (2-17) had relatively pale ventral plumage and formed a very distinct group on axes I and II. The first axis (47.5% of variance) primarily reflected differences in dominant wavelength and brightness, while the second axis represented variation in purity (34.0% of variance). Breast color also was relatively bright in birds from the Eastern Mojave (37), Southern Idaho (41), and North-central Utah (42), which likewise had the brightest dorsal color. Additional samples that showed relatively pale breast color included Joshua Tree (24) and Tejon (18), which was more similar to Santa Barbara (17) than to other samples from southern California.

Cluster analysis. UPGMA analysis identified two major clusters based on all 9 color variables (Fig. 33): Pacific slope samples from Medford (1) to the Sierra de la Laguna (30), plus samples from the San Francisco Mountains (53) and Eastern Arizona (55); and interior samples from the western Great Basin (33, 36), Southern Idaho (41), and Central Utah (43) to the Guadalupe Mountains (60). Birds from the Eastern Mojave (37) and North-central Utah (42) branched outside of the other two groups. Within each cluster, numerous subgroups were recognized that had varying degrees of geographic structure. For example, samples from Ventura (19) to the Sierra de la Laguna (30) grouped together, although the Sierra de la Laguna sample also allied to the San Francisco Mountains (53).

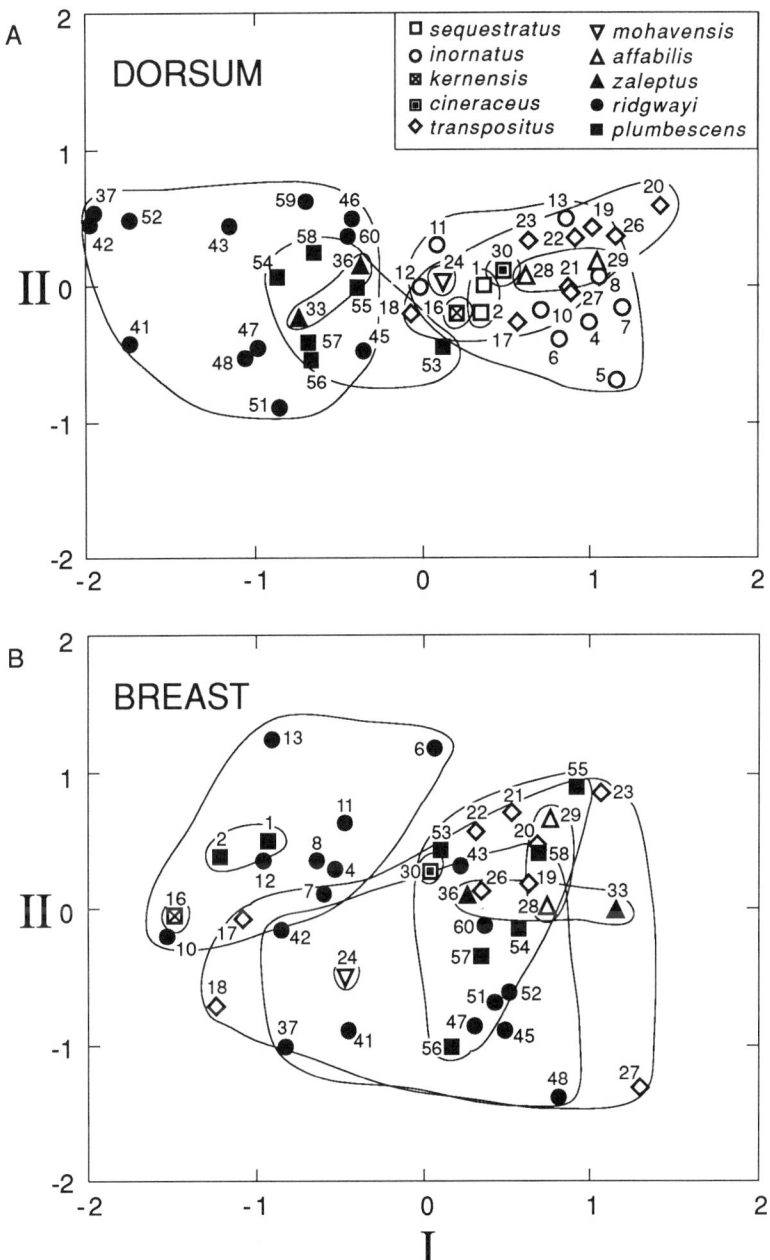

FIGURE 32. Multidimensional scaling analysis of samples based on variation in color of the dorsum (A) and breast (B). Symbols for different subspecies are the same as in Figs. 21-22; sample areas are numbered as in Figs. 3-4 and Table 2. The plot in Fig. 32A achieved a minimum "stress" value of 0.087 after 15 iterations, indicating good to excellent agreement with the original Euclidean distance matrix (Rohlf 1987). Data on breast color had only a fair goodness-of-fit between the MDS plot and distance matrix ("stress" = 0.269 after 16 iterations).

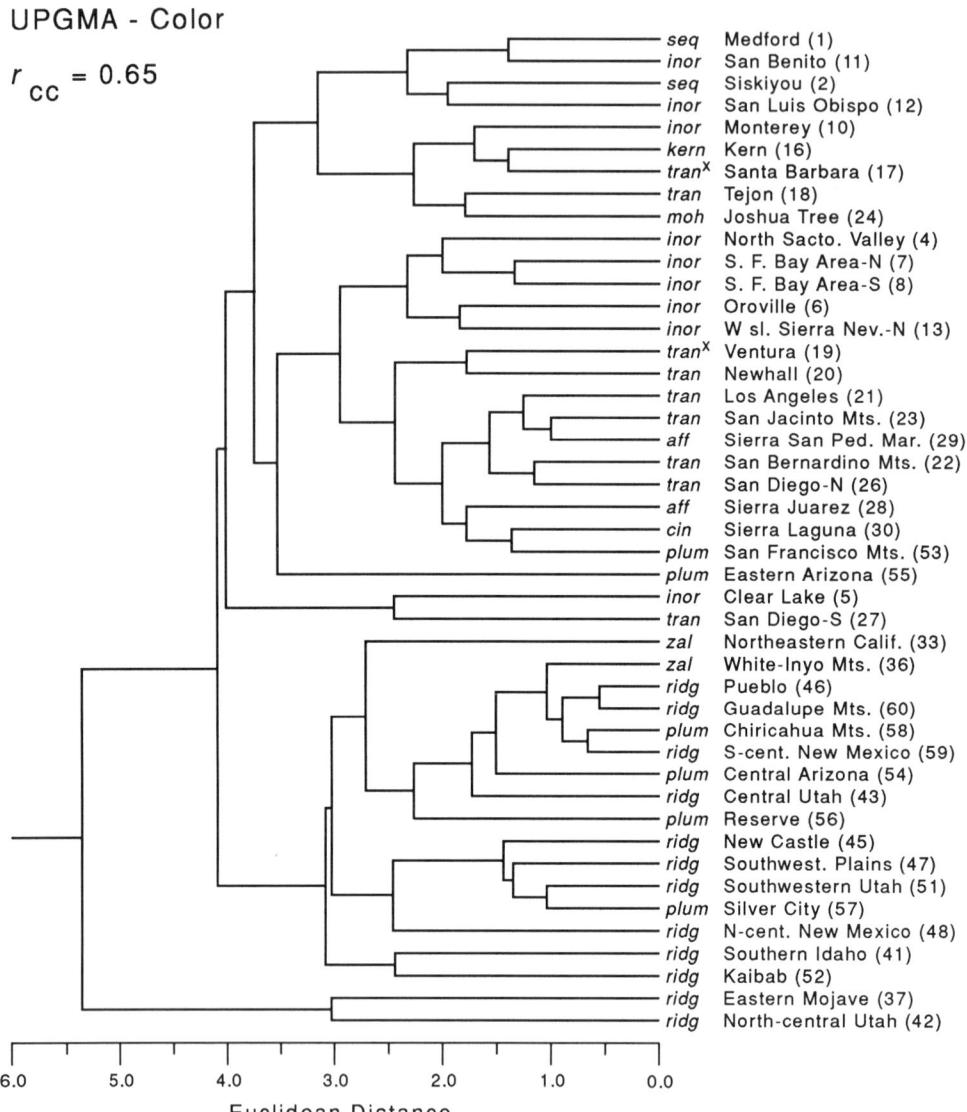

UPGMA - Color

$r_{cc} = 0.65$

seq	Medford (1)
inor	San Benito (11)
seq	Siskiyou (2)
inor	San Luis Obispo (12)
inor	Monterey (10)
kern	Kern (16)
tran[X]	Santa Barbara (17)
tran	Tejon (18)
moh	Joshua Tree (24)
inor	North Sacto. Valley (4)
inor	S. F. Bay Area-N (7)
inor	S. F. Bay Area-S (8)
inor	Oroville (6)
inor	W sl. Sierra Nev.-N (13)
tran[X]	Ventura (19)
tran	Newhall (20)
tran	Los Angeles (21)
tran	San Jacinto Mts. (23)
aff	Sierra San Ped. Mar. (29)
tran	San Bernardino Mts. (22)
tran	San Diego-N (26)
aff	Sierra Juarez (28)
cin	Sierra Laguna (30)
plum	San Francisco Mts. (53)
plum	Eastern Arizona (55)
inor	Clear Lake (5)
tran	San Diego-S (27)
zal	Northeastern Calif. (33)
zal	White-Inyo Mts. (36)
ridg	Pueblo (46)
ridg	Guadalupe Mts. (60)
plum	Chiricahua Mts. (58)
ridg	S-cent. New Mexico (59)
plum	Central Arizona (54)
ridg	Central Utah (43)
plum	Reserve (56)
ridg	New Castle (45)
ridg	Southwest. Plains (47)
ridg	Southwestern Utah (51)
plum	Silver City (57)
ridg	N-cent. New Mexico (48)
ridg	Southern Idaho (41)
ridg	Kaibab (52)
ridg	Eastern Mojave (37)
ridg	North-central Utah (42)

6.0 5.0 4.0 3.0 2.0 1.0 0.0

Euclidean Distance

FIGURE 33. UPGMA phenogram of Euclidean distances separating 45 samples of the *Parus inornatus* complex based on color variation in males and females combined. The cophenetic correlation coefficient (r_{cc}) indicates fair agreement with the original distance matrix. See legend to Fig. 23 for more detail.

Santa Barbara (17) and Tejon (18) clustered distantly from these southern coastal samples, linking to Kern (16) and Joshua Tree (24), respectively. In the interior, subclusters often were comprised of samples from distant areas. Thus, for example, the White-Inyo Mountains (36) grouped with Pueblo (46) and samples from the Southwest (58-60) rather than with those from Northeastern California (33) or the Eastern Mojave. As in the MDS analysis, subclusters did not fall neatly into subspecies groups.

TABLE 13. Canonical correlation analysis between 26 environmental variables and 9 colorimetric characters for males and females combined. Values indicate the correlation between each character in the two data sets and the first canonical variate. The canonical correlation between the colorimetric and environmental data sets on the first canonical variable is 0.984.

Variable		Variable	
Environmental			
Elevation	0.833	Dec. Temp.-Min.	-0.780
Latitude	-0.060	Dec. Temp.-Mean	-0.695
Longitude	-0.795	Jan. Temp.-Max.	-0.569
April Temp.-Max	-0.227	Jan. Temp.-Min.	-0.773
April Temp.-Min	-0.658	Jan. Temp.-Mean	-0.698
April Temp.-Mean	-0.499	Seasonal Temp. Diff.[a]	0.714
May Temp.-Max.	0.024	April Precip.	-0.604
May Temp.-Min.	-0.495	May Precip.	0.272
May Temp.-Mean	-0.274	June Precip.	0.550
June Temp.-Max.	0.259	August Precip.	0.718
June Temp.-Min.	-0.244	Dec. Precip.	-0.716
June Temp.-Mean	0.017	Jan. Precip.	-0.758
Dec. Temp.-Max.	-0.545	Annual Precip.	-0.524
Colorimetric			
Dominant Wavelength-Dorsum	-0.495	Purity-Breast	-0.282
Brightness-Dorsum	-0.863	Dominant Wavelength-Belly	-0.241
Purity-Dorsum	0.670	Brightness-Belly	-0.324
Dominant Wavelength-Breast	-0.091	Purity-Belly	-0.311
Brightness-Breast	-0.246		

[a]Calculated as the difference between June mean temperature and December mean temperature.

Mantel Tests of Congruence Between Size and Color

Comparisons of Euclidean distance matrices based on size and color data, respectively, revealed a significant association for both males and females: males, $r = 0.274$, $t = 3.795$; females, $r = 0.248$, $t = 3.417$. The t-statistic is significant when values exceed 2.694 ($P < 0.0071$ based on the procedure-wise error rate). In both of these comparisons, the probability was 1.0 that a random Mantel statistic Z is less than the observed Z.

Environmental Correlates of Geographic Variation in Color

Color variation was strongly correlated with environment, especially in terms of brightness and purity of dorsal plumage (Table 13). The environmental factors that contributed most heavily to this relationship were similar to those associated with size variation in titmice, again reflecting differences in climate experienced by Pacific slope and interior populations. Thus, environment appears to exert a similar influence on colorimetric and morphometric variation among titmice in the western United States.

GENETIC VARIATION

ALLOZYME VARIATION

Heterozygosity and Levels of Genetic Variation

Of the 33 loci scored, 16 (48.5%) were variable within the *Parus inornatus* complex (Table 14). *Parus bicolor* and *P. gambeli* did not vary at any locus that was monomorphic in *P. inornatus*. However, 7 alleles occurred in the outgroups that were not observed in a single sample of *P. inornatus* (αGPD, d; 6PGD, d; SOD-1, a; GPT, b; ADA, a; MPI, c; and GPI, e). *Parus bicolor* contained a fixed difference at 2 alleles, whereas *P. gambeli* was fixed at 5 alleles not present in *P. inornatus*. The following loci (17; 51.5%) were monomorphic in all individuals examined (see Table 3 for Enzyme Commission numbers): GDA, AB-1, AB-2, AB-3, AB-4, ADH, SDH, MDH-1, MDH-2, ICD-2, GLUD, GOT-2, EAP, LDH-1, SOD-2, CK-1, and CK-2.

Levels of genetic variability within samples of the *P. inornatus* complex are given in Table 15. Observed heterozygosities ($H_{obs.}$) ranged from 0.006 in the Medford sample (1) to 0.044 in South-central New Mexico (59), averaging 0.025 over all populations. The percentage of polymorphic loci (POLY) varied from 6.06% to 27.27%, while the mean number of alleles per locus (NALL) ranged from 1.06 to 1.30. POLY and NALL averaged 13.81% and 1.18, respectively. All three measures generally revealed lower variability in samples from the Pacific slope than from the interior. Among the Pacific slope samples, $H_{obs.}$ was substantially higher in Clear Lake (5) than in other samples from the same region; POLY and NALL also were highest in that sample. Other fairly variable samples included Monterey (10), Tejon (18), and southern San Diego (27). Although the lowest values were obtained for Medford, Kern (16) and Joshua Tree (24) likewise were relatively invariant. Within the interior, titmice from the Eastern Mojave (37) and Chiricahua Mountains (58) showed strikingly lower variability compared to other samples. Genetic variation also was relatively low in Reno (34) and Benton (35), especially in terms of POLY and NALL.

91

TABLE 14. Allelic frequencies and *F*-statistics at 16 variable loci in 32 samples of the *Parus inornatus* complex and 2 outgroups (*P. bicolor, P. gambeli*). Sample area number is given above name. See Table 3 for Enzyme Commission numbers.

Locus	Allele	(1) Medford	(2) Siskiyou	(3) Mount Dome	(4) N Sacto. Valley	(5) Clear Lake	(6) Oroville	(10) Monterey
αGPD	a							0.156
	b							
	c	1.000	1.000	1.000	1.000	1.000	1.000	0.844
	d							
LDH-2	a	1.000	1.000	1.000	1.000	1.000	1.000	1.000
	b							
ICD-1	a							
	b							
	c	1.000	1.000	1.000	1.000	1.000	1.000	1.000
	d							
6-PGD	a							
	b	1.000	1.000	1.000	1.000	0.941	0.964	1.000
	c					0.059		
	d							
	e						0.036	
SOD-1	a							
	b	1.000	1.000	1.000	1.000	1.000	1.000	1.000
	c							
NP	a				0.031		0.071	
	b	0.033	0.097	0.060		0.029		
	c		0.016		0.031	0.147		
	d	0.967	0.887	0.940	0.938	0.824	0.929	1.000
	e							
GOT-1	a						0.036	
	b	1.000	1.000	1.000	1.000	1.000	0.964	1.000
	c							
	d							
GPT	a							
	b							
	c	1.000	1.000	1.000	0.938	1.000	1.000	1.000
	d							
	e				0.063			

TABLE 14 (continued)

(12) San Luis Obispo	(15) W sl. Sierra Nevada-S	(16) Kern	(18) Tejon	(24) Joshua Tree	(27) San Diego-S	(31) Warner	(32) Tule Lake	(33) NE Calif.
0.094				0.050				
0.906	1.000	1.000	1.000	0.950	1.000	1.000	1.000	1.000
1.000	1.000	1.000	1.000	1.000	1.000	1.000	1.000	1.000
							0.038	0.028
1.000	0.967	1.000	1.000	1.000	1.000	1.000	0.962	0.972
	0.033							
					0.029			
1.000	1.000	0.967	0.912	1.000	0.941	1.000	1.000	1.000
		0.033	0.088		0.029			
1.000	1.000	1.000	1.000	1.000	1.000	1.000	1.000	1.000
		0.033	0.029					
		0.033		0.050				
0.063	0.067		0.118	0.100	0.059			
0.938	0.933	0.933	0.853	0.850	0.941	1.000	1.000	1.000
	0.033		0.029					
1.000	0.967	1.000	0.971	1.000	0.941	1.000	1.000	1.000
					0.059			
1.000	1.000	1.000	1.000	1.000	1.000	1.000	1.000	1.000

TABLE 14 (continued)

Locus	Allele	(1) Medford	(2) Siskiyou	(3) Mount Dome	(4) N Sacto. Valley	(5) Clear Lake	(6) Oroville	(10) Monterey
CK-3	a							
	b	1.000	1.000	1.000	1.000	1.000	1.000	1.000
PGM-1	a	0.067						
	b	0.933	1.000	1.000	1.000	1.000	1.000	1.000
	c							
EST-D	a					0.029		0.031
	b	1.000	1.000	1.000	1.000	0.971	1.000	0.969
	c							
LA-1	a							
	b	1.000	1.000	1.000	0.969	1.000	0.964	0.969
	c				0.031		0.036	0.031
	d							
LGG	a							
	b			0.080	0.031	0.147	0.036	0.063
	c	1.000	0.887	0.740	0.969	0.735	0.929	0.938
	d		0.113	0.180		0.118		
	e						0.036	
ADA	a							
	b							
	c	1.000	1.000	1.000	1.000	1.000	1.000	1.000
	d							
MPI	a					0.029		
	b	1.000	1.000	1.000	1.000	0.971	1.000	1.000
	c							
	d							
GPI	a							
	b			0.020				0.031
	c							0.031
	d	1.000	1.000	0.980	1.000	1.000	1.000	0.938
	e							

TABLE 14 (continued)

(12) San Luis Obispo	(15) W sl. Sierra Nevada-S	(16) Kern	(18) Tejon	(24) Joshua Tree	(27) San Diego-S	(31) Warner	(32) Tule Lake	(33) NE Calif.
0.031								
0.969	1.000	1.000	1.000	1.000	1.000	1.000	1.000	1.000
					0.147	0.188		
1.000	1.000	1.000	0.971	1.000	0.853	0.813	1.000	1.000
			0.029					
		0.033						0.028
1.000	1.000	0.967	1.000	1.000	1.000	1.000	1.000	0.972
0.969	0.967	1.000	0.941	1.000	0.971	1.000	1.000	1.000
0.031	0.033		0.059		0.029			
						0.063	0.192	0.222
1.000	1.000	1.000	1.000	1.000	1.000	0.438	0.615	0.333
						0.500	0.192	0.444
								0.028
1.000	1.000	1.000	1.000	1.000	1.000	1.000	1.000	0.972
1.000	1.000	1.000	1.000	1.000	1.000	1.000	1.000	1.000
						0.125	0.115	0.222
1.000	1.000	1.000	1.000	1.000	1.000	0.875	0.885	0.778

TABLE 14 (continued)

Locus	Allele	(34) Reno	(35) Benton	(36) White-Inyo Mts.	(37) Eastern Mojave	(39) SE Nevada	(40) Ruby Mts.	(42) N-cent. Utah	(44) La Sal
αGPD	a								
	b						0.067		
	c	1.000	1.000	1.000	1.000	1.000	0.933	1.000	1.000
	d								
LDH-2	a	1.000	1.000	1.000	1.000	1.000	1.000	1.000	1.000
	b								
ICD-1	a			0.033					
	b								
	c	1.000	1.000	0.967	1.000	1.000	1.000	1.000	1.000
	d								
6-PGD	a								
	b	1.000	1.000	1.000	0.967	1.000	1.000	1.000	0.938
	c				0.033				0.063
	d								
	e								
SOD-1	a								
	b	1.000	1.000	1.000	1.000	1.000	0.967	1.000	1.000
	c						0.033		
NP	a								
	b								
	c							0.026	
	d	1.000	1.000	1.000	1.000	1.000	1.000	0.974	1.000
	e								
GOT-1	a								
	b	1.000	1.000	1.000	1.000	1.000	1.000	1.000	1.000
	c								
	d								
GPT	a					0.056			
	b								
	c	1.000	1.000	1.000	1.000	0.917	1.000	1.000	1.000
	d					0.028			
	e								

TABLE 14 (continued)

(45) New Castle	(48) N-cent. New Mex.	(49) Zuni Mts.	(52) Kaibab	(53) San Fran. Mts.	(56) Reserve	(58) Chiricahua Mts.	(59) S-cent. New Mex.
1.000	1.000	1.000	1.000	1.000	1.000	1.000	1.000
1.000	1.000	1.000	1.000	1.000	0.967	1.000	1.000
					0.033		
					0.033		
1.000	1.000	1.000	1.000	1.000	0.967	1.000	1.000
0.938	1.000	0.971	1.000	1.000	1.000	1.000	0.969
0.063		0.029					0.031
1.000	1.000	1.000	1.000	1.000	1.000	1.000	1.000
		0.029		0.059	0.100		0.094
0.961	1.000	0.971	1.000	0.941	0.900	1.000	0.906
0.031							
			0.033				0.031
1.000	1.000	0.971	0.967	1.000	1.000	1.000	0.969
		0.029					
	0.050	0.029			0.100		
1.000	0.950	0.912	1.000	1.000	0.900	0.958	1.000
		0.059				0.042	

TABLE 14 (continued)

Locus	Allele	(34) Reno	(35) Benton	(36) White-Inyo Mts.	(37) Eastern Mojave	(39) SE Nevada	(40) Ruby Mts.	(42) N-cent. Utah	(44) La Sal
CK-3	a								
	b	1.000	1.000	1.000	1.000	1.000	1.000	1.000	1.000
PGM-1	a		0.033	0.100		0.083		0.026	
	b	1.000	0.967	0.900	1.000	0.889	1.000	0.974	1.000
	c					0.028			
EST-D	a					0.056	0.033		
	b	1.000	1.000	1.000	1.000	0.944	0.967	0.974	1.000
	c							0.026	
LA-1	a								
	b	1.000	1.000	1.000	1.000	0.972	1.000	1.000	0.938
	c					0.028			0.063
	d								
LGG	a							0.079	
	b	0.077	0.100	0.267		0.222	0.267	0.132	0.313
	c	0.615	0.567	0.633	0.867	0.556	0.533	0.763	0.688
	d	0.308	0.333	0.100	0.133	0.222	0.200	0.026	
	e								
ADA	a								
	b			0.067	0.033				
	c	1.000	1.000	0.933	0.967	1.000	1.000	1.000	1.000
	d								
MPI	a								
	b	1.000	1.000	1.000	1.000	0.944	0.967	0.947	0.969
	c								
	d					0.056	0.033	0.053	0.031
GPI	a			0.067					
	b	0.038	0.067			0.083	0.033	0.211	0.156
	c								
	d	0.962	0.933	0.933	1.000	0.917	0.967	0.789	0.844
	e								

TABLE 14 (continued)

(45) New Castle	(48) N-cent. New Mex.	(49) Zuni Mts.	(52) Kaibab	(53) San Fran. Mts.	(56) Reserve	(58) Chiricahua Mts.	(59) S-cent. New Mex.
1.000	1.000	1.000	1.000	1.000	1.000	1.000	1.000
1.000	1.000	1.000	1.000	1.000	1.000	1.000	1.000
	0.075	0.029		0.029	0.033		
1.000	0.925	0.971	0.967	0.941	0.967	1.000	1.000
			0.033	0.029			
		0.029					
0.875	0.925	0.971	1.000	0.912	0.933	1.000	0.906
0.125	0.075			0.088	0.067		0.063
							0.031
0.031	0.100	0.029	0.033	0.059	0.067	0.083	0.094
0.125	0.125	0.971	0.133	0.235	0.200	0.083	0.219
0.813	0.650		0.833	0.647	0.700	0.750	0.656
0.031	0.125			0.059	0.033	0.083	0.031
		0.176					
1.000	1.000	0.824	1.000	1.000	1.000	1.000	0.906
							0.094
	0.025						
1.000	0.975		0.967	0.912	1.000	1.000	1.000
		1.000					
			0.033	0.088			
		0.033					
0.156		1.000	0.167	0.088	0.100	0.042	0.156
	0.025			0.029			
0.844	0.975		0.800	0.882	0.900	0.958	0.844

TABLE 14 (continued)

Locus	Allele	Parus bicolor	Parus gambeli	Parus inornatus complex		
				F_{IS}	F_{IT}	F_{ST}
αGPD	a			0.068	0.162	0.101
	b					
	c	1.000				
	d		1.000			
LDH-2	a	1.000	1.000	-0.035	-0.001	0.032
	b					
ICD-1	a			-0.035	-0.004	0.030
	b					
	c	1.000	1.000			
	d					
6-PGD	a			-0.058	-0.014	0.042
	b	1.000				
	c					
	d		1.000			
	e					
SOD-1	a	1.000		-0.035	-0.001	0.032
	b		1.000			
	c					
NP	a			-0.053	0.010	0.059
	b					
	c		1.000			
	d	1.000				
	e					
GOT-1	a			-0.040	-0.006	0.033
	b	0.833	1.000			
	c					
	d	0.167				
GPT	a			-0.075	-0.009	0.061
	b	1.000	1.000			
	c					
	d					
	e					

TABLE 14 (continued)

Locus	Allele	Parus bicolor	Parus gambeli	Parus inornatus complex F_{IS}	F_{IT}	F_{ST}
CK-3	a			-0.032	-0.001	0.030
	b	1.000	1.000			
PGM-1	a			-0.131	-0.022	0.097
	b	1.000	1.000			
	c					
EST-D	a			-0.045	-0.012	0.031
	b	1.000	1.000			
	c					
LA-1	a			-0.073	-0.024	0.045
	b	1.000				
	c		1.000			
	d					
LGG	a			-0.069	0.102	0.160
	b		0.100			
	c	1.000	0.900			
	d					
	e					
ADA	a		1.000	-0.073	-0.005	0.063
	b					
	c	1.000				
	d					
MPI	a			-0.055	-0.010	0.044
	b	1.000				
	c		1.000			
	d					
GPI	a			-0.026	0.058	0.081
	b					
	c					
	d	0.833				
	e	0.167	1.000			
Mean				-0.058	0.054	0.106

TABLE 15. Genetic variability measures for 32 samples of the *Parus inornatus* complex. Sample areas are numbered as in Figs. 3-4 and Table 2; see Table 2 for sample sizes.

Sample Area	$H_{obs.} \pm$ SE	$H_{exp.} \pm$ SE[a]	POLY[b]	NALL[c]
1. Medford	0.006 ± 0.004	0.006 ± 0.004	6.06	1.06
2. Siskiyou	0.012 ± 0.008	0.012 ± 0.009	6.06	1.09
3. Mt. Dome	0.017 ± 0.013	0.017 ± 0.013	9.09	1.12
4. North Sacramento Valley	0.011 ± 0.006	0.011 ± 0.006	12.12	1.15
5. Clear Lake	0.029 ± 0.015	0.030 ± 0.016	15.15	1.21
6. Oroville	0.015 ± 0.007	0.015 ± 0.007	15.15	1.18
10. Monterey	0.017 ± 0.008	0.019 ± 0.010	15.15	1.18
12. San Luis Obispo	0.013 ± 0.007	0.013 ± 0.007	12.12	1.12
15. W slope Sierra Nevada-S	0.010 ± 0.005	0.010 ± 0.005	12.12	1.12
16. Kern	0.008 ± 0.005	0.008 ± 0.005	9.09	1.12
18. Tejon	0.018 ± 0.008	0.020 ± 0.010	15.15	1.18
24. Joshua Tree	0.012 ± 0.009	0.011 ± 0.009	6.06	1.09
27. San Diego-South	0.021 ± 0.010	0.020 ± 0.010	15.15	1.18
31. Warner	0.038 ± 0.023	0.034 ± 0.020	9.09	1.12
32. Tule Lake	0.028 ± 0.020	0.026 ± 0.018	9.09	1.12
33. Northeastern California	0.037 ± 0.024	0.036 ± 0.022	15.15	1.18
34. Reno	0.026 ± 0.023	0.019 ± 0.017	6.06	1.09
35. Benton	0.022 ± 0.017	0.023 ± 0.018	9.09	1.12
36. White-Inyo Mts.	0.036 ± 0.021	0.032 ± 0.018	15.15	1.18
37. Eastern Mojave	0.012 ± 0.008	0.011 ± 0.008	9.09	1.09
39. Southeastern Nevada	0.040 ± 0.018	0.043 ± 0.020	21.21	1.30
40. Ruby Mts.	0.034 ± 0.022	0.031 ± 0.019	18.18	1.21
42. North-central Utah	0.032 ± 0.017	0.030 ± 0.016	18.18	1.24
44. La Sal	0.034 ± 0.018	0.031 ± 0.016	15.15	1.15
45. New Castle	0.030 ± 0.015	0.031 ± 0.015	14.71	1.21
48. N-central New Mexico	0.035 ± 0.020	0.031 ± 0.018	18.18	1.24
49. Zuni	0.029 ± 0.010	0.030 ± 0.011	27.27	1.30

TABLE 15 (continued)

Sample Area	$H_{obs.} \pm$ SE	$H_{exp.} \pm$ SE[a]	POLY[b]	NALL[c]
52. Kaibab	0.022 ± 0.011	0.025 ± 0.014	15.15	1.21
53. San Francisco Mts.	0.043 ± 0.020	0.040 ± 0.019	18.18	1.30
56. Reserve	0.042 ± 0.017	0.041 ± 0.017	24.24	1.30
58. Chiricahua Mts.	0.015 ± 0.011	0.018 ± 0.013	9.09	1.15
59. S-central New Mexico	0.044 ± 0.018	0.044 ± 0.019	21.21	1.30
Mean	0.025	0.024	13.81	1.18

[a] Unbiased estimate (Nei 1978).
[b] Percentage of loci polymorphic (frequency of most common allele $\leq$ 0.99).
[c] Mean number of alleles per locus.

Intra-population heterozygosity was significantly correlated ($P < 0.01$) with the percentage of polymorphic loci ($r = 0.654$) and the mean number of alleles per locus ($r = 0.733$). Furthermore, POLY and NALL were very highly correlated ($r = 0.940$). Correlations between sample size and each of these three measures of genetic variability showed no association (n vs. $H_{obs.}$, $r = 0.020$; n vs. POLY, $r = 0.028$; n vs. NALL, $r = 0.052$). The lack of a significant relationship between sample size and genetic variability is supported by previous theoretical (Nei 1978) and empirical (Gorman and Renzi 1979) work, given the large number of loci surveyed in this study.

Chi-square tests indicated that samples did not deviate significantly ($P > 0.05$) from Hardy-Weinberg expectations at all variable loci examined. One exception involved variability at LGG in the Chiricahua Mountains (58), where the observed heterozygote frequencies were greater than those expected based on Hardy-Weinberg equilibrium ($X^2 = 23.48$, $P = 0.001$). However, the chi-square test for goodness-of-fit may be suspect if the expected frequencies of some classes are low (Sokal and Rohlf 1969; Swofford and Selander 1981). Heterozygosity at LGG in the Chiricahua Mountains was similar to that expected under Hardy-Weinberg equilibrium when genotypes were pooled into three classes ($X^2 = 0.307$, $P = 0.580$; see Swofford and Selander 1981:17).

Because levels of allozymic variability clearly differed geographically, I ran a canonical correlation analysis to assess whether the pattern of variation could be attributed to environmental factors (Table 16). The three canonical variates reflected different measures of genetic variability, with observed heterozygosity (second canonical variate) showing the highest correlation with environmental variables. However, unlike the results of the morphometric and colorimetric analysis, no environmental factors emerged as being overwhelmingly important in this analysis.

TABLE 16. Canonical-correlation analysis between 26 environmental variables and allozymic variability. Values indicate the correlation between each character in the two data sets and three canonical variates. Canonical correlations between the allozymic and environmental data sets are 0.973 for the first canonical variate, 0.966 for the second variate, and 0.871 for the third variate.

Variable	Canonical Variate			Variable	Canonical Variate		
	1	2	3		1	2	3
Environmental							
Elevation	-0.010	-0.667	0.358	Dec. Temp.-Min.	-0.102	0.682	-0.296
Latitude	0.401	-0.116	-0.311	Dec. Temp.-Mean	-0.169	0.639	-0.180
Longitude	0.329	0.414	-0.505	Jan. Temp.-Max.	-0.191	0.557	-0.054
April Temp.-Max	-0.147	0.431	0.090	Jan. Temp.-Min.	-0.104	0.676	-0.298
April Temp.-Min	-0.150	0.592	-0.246	Jan. Temp.-Mean	-0.151	0.643	-0.188
April Temp.-Mean	-0.158	0.557	-0.097	Seasonal Temp. Diff.[a]	0.115	-0.439	0.238
May Temp.-Max.	-0.072	0.360	0.148	April Precip.	-0.256	0.413	-0.061
May Temp.-Min.	-0.134	0.504	-0.200	May Precip.	0.045	-0.246	-0.038
May Temp.-Mean	-0.112	0.467	-0.037	June Precip.	0.171	-0.432	0.138
June Temp.-Max.	-0.053	0.236	0.208	August Precip.	-0.186	-0.230	0.512
June Temp.-Min.	-0.125	0.423	-0.120	Dec. Precip.	-0.064	0.621	-0.068
June Temp.-Mean	-0.096	0.357	0.042	Jan. Precip.	-0.093	0.610	-0.050
Dec. Temp.-Max.	-0.225	0.525	-0.028	Annual Precip.	-0.211	0.532	0.130
Allozymic							
Heterozygosity	0.119	-0.815	0.567				
POLY	-0.713	-0.530	0.460				
NALL	-0.412	-0.384	0.827				

[a] Calculated as the difference between June mean temperature and December mean temperature.

Geographic Trends in Allelic Frequencies

A contingency table analysis of inter-population heterogeneity (Swofford and Selander 1981:38-40) revealed strong geographic differences at all loci combined. Of the 16 polymorphic loci, 8 (αGPD, NP, GOT-1, GPT, PGM-1, LGG, ADA, GPI) were significantly variable among populations ($X^2 = 119.1$-408.2, $P < 0.05$). Chi-square tests were not significant for the other 8 loci.

Three loci exhibited particularly clear geographic patterns in the distribution and frequencies of alleles (Figs. 34 and 35). LGG (Fig. 34B) was most variable among samples ($X^2 = 408.2$) and exhibited the highest F_{ST} value (0.160) of any locus scored (Table 14). A total of 5 alleles was recorded at LGG in the 32 samples of *P. inornatus*, including one unique allele (LGG-e) found only in the Oroville (6) sample. The fastest allele (LGG-a) occurred at low to moderate frequency (3.1-10.0%) from North-central Utah (42) to South-central New Mexico (59), but was absent in samples from La Sal (44) and the Zuni Mountains (49); samples from eastern Nevada (39-40) to the Pacific coast all lacked this allele. Allele b was present in moderate to high frequency (6.3-31.3%) in every interior sample with the exception of the Eastern Mojave (37). This allele also occurred in fairly high frequency at Mount Dome (3; 8.0%) and Clear Lake (5; 14.7%), and showed up occasionally in birds from the North Sacramento Valley (4; 3.1%), Oroville (6; 3.6%), and Monterey (10; 6.3%). Pacific slope samples from Medford (1) and from central to southern California (12-27) were monomorphic for the most common allele (LGG-c). Although allele c had the highest frequency in all samples except those from Warner (31; 43.8%) and Northeastern California (33; 33.3%), frequencies of this allele were noticeably lower in samples from the western Great Basin (31 to 36) and eastern Nevada (39-40). These samples likewise exhibited a higher frequency of the slow allele d compared to other populations. While the Siskiyou (2) sample lacked allele b, in contrast to Mt. Dome (3), both samples shared allele d at moderate frequency. The frequency of allele d increased slightly from 11.3% at Siskiyou to 18.0% at Mount Dome and 19.2% at Tule Lake (32), then rose sharply to 50.0% and 44.4% at Warner (31) and in Northeastern California (33), respectively. High frequencies of this slow allele also occurred southward to Reno (34; 30.8%) and Benton (35; 33.3%), and eastward to Southeastern Nevada (39; 22.2%) and the Ruby Mountains (40; 20.0%). On the other hand, frequencies were lower in samples from the White-Inyo Mountains (36; 10.0%), Eastern Mojave (37; 13.3%), and other interior populations (2.6-12.5%). Allele d was absent from three interior samples (La Sal [44], Zuni Mountains [49], Kaibab [52]) and from all Pacific slope samples except for Clear Lake (5), where it occurred in moderate frequency (11.8%).

Genetic differentiation between Pacific slope and interior samples also was evident at the GPI locus (Fig. 35A), which had an overall F_{ST} of 0.081. Four alleles were recorded in *P. inornatus*, two of which were restricted to a few samples. Titmice from the White-Inyo Mountains (36) and Kaibab (52) shared an allele (GPI-a) at frequencies of

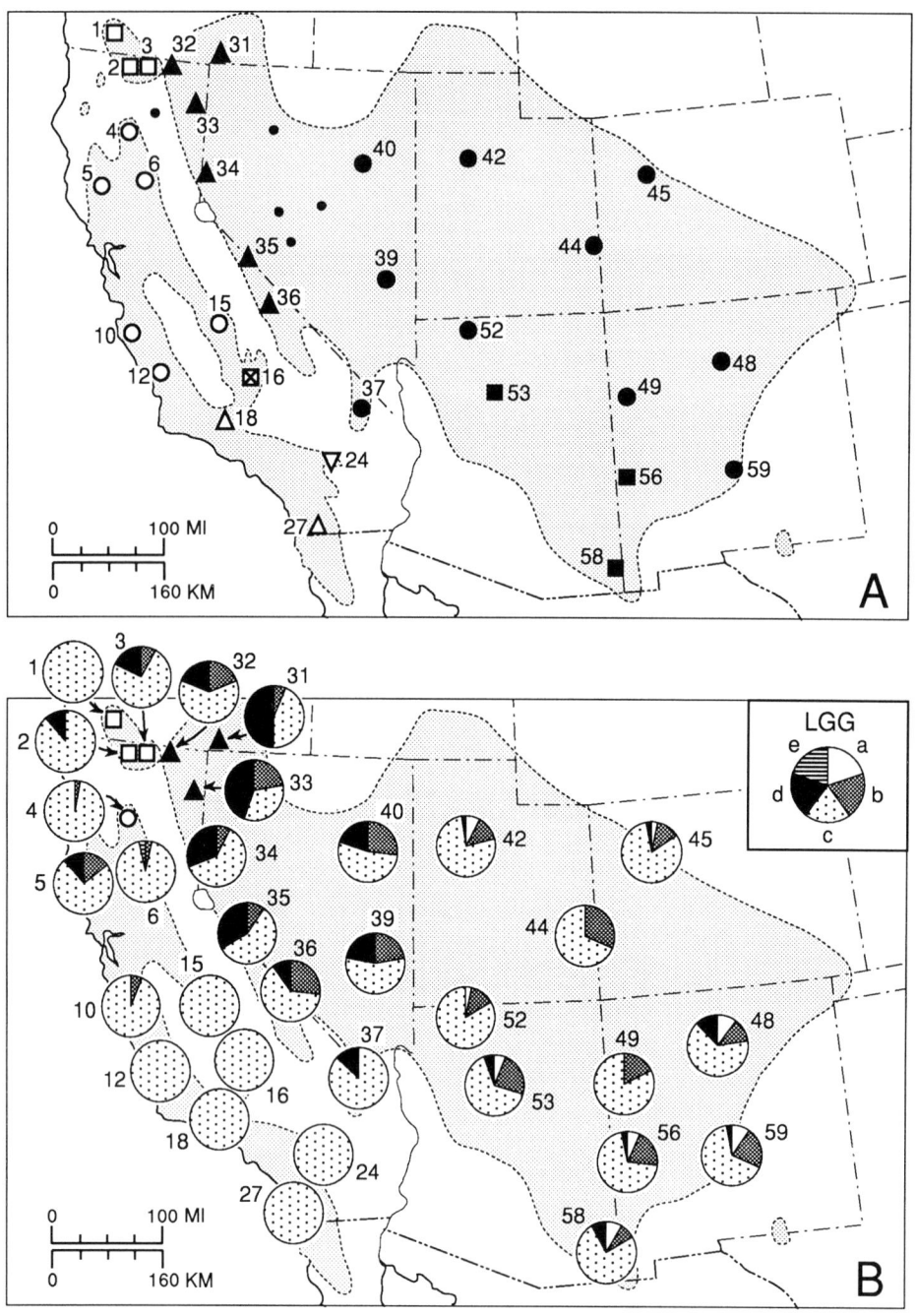

FIGURE 34. Geographic distribution of tissue samples (A) and frequencies of alleles at the LGG locus (B) for 32 samples of the *Parus inornatus* complex. Symbols in Fig. 34A represent subspecies recognized by the American Ornithologists' Union (1957); see Figs. 21-22 and Fig. 32. Dots indicate additional tissue samples not analyzed allozymically. Samples are numbered as in Figs. 3-4 and Table 2.

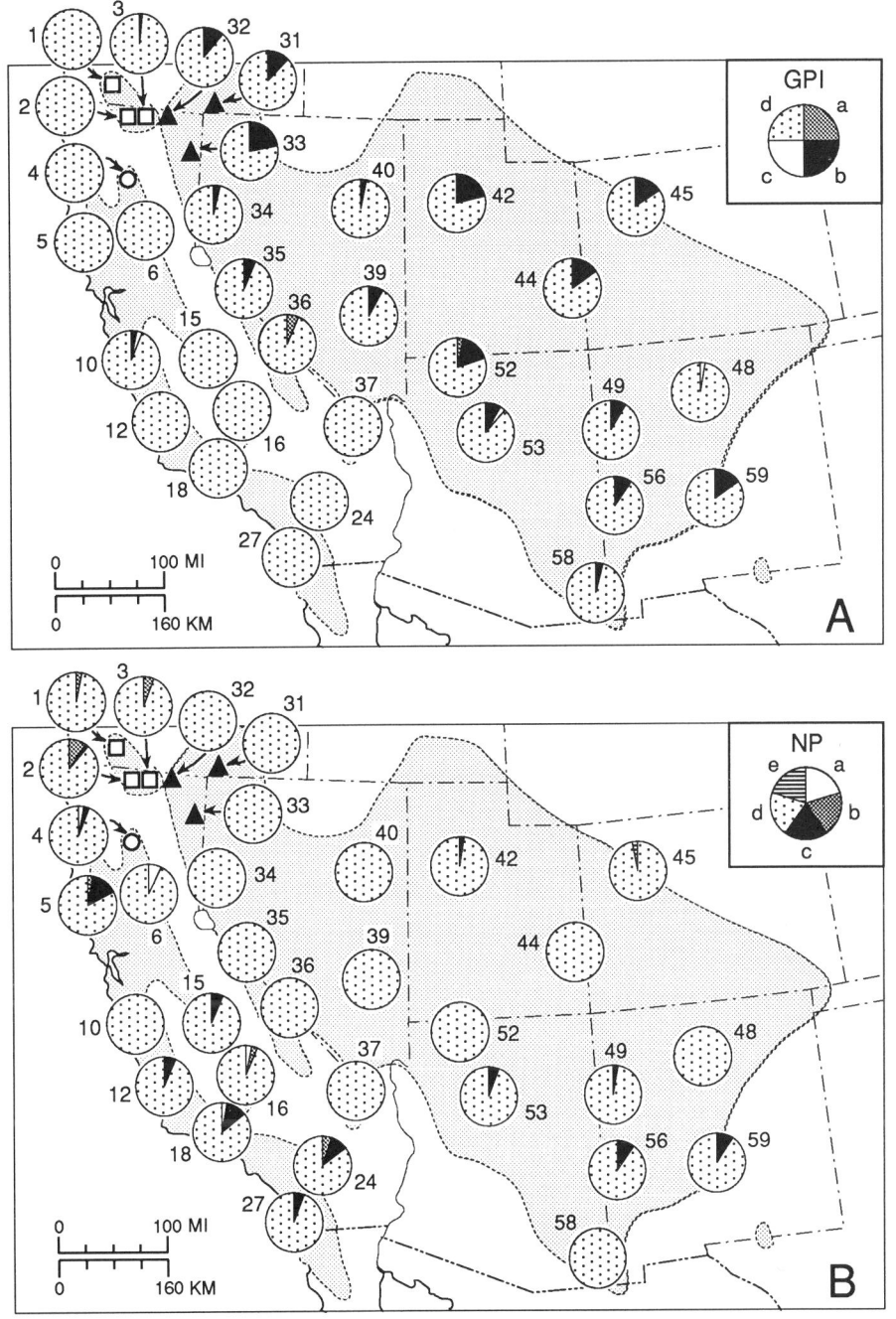

FIGURE 35. Geographic distribution and frequencies of alleles at the GPI (A) and NP (B) loci for 32 samples of the *Parus inornatus* complex. See legend to Fig. 34 for more detail.

6.7% and 3.3%, respectively; this allele was absent in all other samples. Similarly, allele c was found only in samples from Monterey (10; 3.1%), North-central New Mexico (48; 2.5%), and the San Francisco Mountains (53; 2.9%). Allele b showed the strongest geographic structure, occurring in most interior samples at frequencies ranging from 3.3% (Ruby Mountains, 40) to 22.2% (Northeastern California, 33). It also appeared in low frequency at Mount Dome (3; 2.0%), but was absent from the White-Inyo Mountains (36), Eastern Mojave (37), and North-central New Mexico (48). The Eastern Mojave sample (37) was monomorphic for the most common allele d, as were all Pacific slope samples with the exception of Monterey (10). Monterey had three alleles (b, c, d), two of which (b, c) occurred at equally low frequencies of 3.1%.

The NP locus ($F_{ST} = 0.059$) exhibited a slightly more complicated pattern (Fig. 35B) than either LGG or GPI. In contrast to LGG and GPI, variability tended to be higher among Pacific slope samples compared to those from the interior. Of the 5 alleles scored, 2 were found only in samples from the Pacific slope region. Four of these samples (North Sacramento Valley [4], Oroville [6], Kern [16], Tejon [18]) shared the fastest allele (NP-a) at frequencies ranging from 2.9% to 7.1%. Another fast allele (NP-b) occurred in 6 Pacific slope samples from Medford (1; 3.3%) to Joshua Tree (24; 5.0%). Samples from Siskiyou (2) and Mount Dome (3) shared this allele in relatively high frequency (9.7% and 6.0%, respectively). Importantly, allele b was absent from all interior samples, including Tule Lake (32). Samples from the western Great Basin to the Eastern Mojave and Ruby Mountains (31-40) were monomorphic for the most common allele d. Most other interior samples (42-59) either were monomorphic for allele d or contained a fast allele (NP-c) at low to moderate frequency (2.6-10.0%). This allele also occurred in 8 Pacific slope samples at frequencies ranging from 1.6% (Siskiyou, 2) to 14.7% (Clear Lake, 5). A unique slow allele (NP-e) was recorded uncommonly at New Castle (45; 3.1%).

Four other loci had relatively high F_{ST} values (Table 14) and warrant discussion. At αGPD ($F_{ST} = 0.101$), Monterey (10), San Luis Obispo (12), and Joshua Tree (24) shared a fast allele (αGPD-a) that was absent in all other samples; this allele occurred at moderately high frequency (15.6%) in Monterey. In addition, the Ruby Mountains had a unique allele (αGPD-b) with a frequency of 6.7%. GPT ($F_{ST} = 0.061$) contained two alleles (a, d) that were present only in the interior, as well as a third allele (GPT-e) unique to the Mount Dome sample (3; 6.3%). PGM-1 had one of the highest F_{ST} values (0.097) of all loci scored, but showed a complex pattern. Perhaps the most notable difference at this locus was the high frequency of allele a in samples from Warner (31; 18.8%) and southern San Diego (27; 14.7%). Finally, samples from Northeastern California (33), the White-Inyo Mountains (36), and the Eastern Mojave (37) were polymorphic for a fast allele (b) at the ADA locus ($F_{ST} = 0.063$). A unique allele at this locus (ADA-c) was found in South-central New Mexico (59).

Genetic Distances

Nei's (1978) genetic distances within the *Parus inornatus* complex ranged from 0.000 to 0.013, averaging 0.00214 ± 0.00242 (Table 17). Inter-population Nei's D was lower among samples from the Pacific slope region ($\bar{D} = 0.00063 \pm 0.00076$) compared to the interior ($\bar{D} = 0.00146 \pm 0.00183$); higher distances resulted from pairwise comparisons between regions ($\bar{D} = 0.00327 \pm 0.00277$). The largest distances ($D = 0.005$ to 0.013, $\bar{D} = 0.00996 \pm 0.00203$) occurred between samples from the Pacific slope (Mount Dome [3] excluded) and those from Warner (31) or Northeastern California (33). These two populations also were fairly distant genetically ($D = 0.004$ to 0.007, $\bar{D} = 0.00540 \pm 0.00099$) to samples ranging from North-central Utah (42) to South-central New Mexico (59). Nei's D was 0.000 between Mount Dome and both Siskiyou (2) and Tule Lake (32), but was relatively high (0.004 to 0.005) when comparing Mount Dome to Warner and Northeastern California; the distance between Medford (1) and Mount Dome was 0.002. Overall, Mount Dome was slightly closer genetically to interior samples ($\bar{D} = 0.00094 \pm 0.001162$) than to those from the Pacific slope ($\bar{D} = 0.00142 \pm 0.000793$), although the variance was higher. All pairwise comparisons with Tule Lake revealed a closer affinity to samples from the interior ($D = 0.000$ to 0.002, $\bar{D} = 0.00044 \pm 0.00070$) than to those from the Pacific slope ($D = 0.001$ to 0.004, $\bar{D} = 0.003 \pm 0.00095$). Surprisingly, birds from the Eastern Mojave (37) were genetically closer to those from the Pacific slope ($D = 0.000$ to 0.001, $\bar{D} = 0.00050 \pm 0.00052$) than to those from the interior ($D = 0.001$ to 0.008, $\bar{D} = 0.00217 \pm 0.00192$); inter-population distances were especially high between the Eastern Mojave and Warner ($D = 0.006$) or Northeastern California ($D = 0.008$).

Nei's distance was much lower between samples of *Parus inornatus* and *P. bicolor* ($\bar{D} = 0.06556 \pm 0.00307$) than between *P. inornatus* and *P. gambeli* ($\bar{D} = 0.27434 \pm 0.00658$). In a previous electrophoretic study of the genus *Parus* (Gill et al. 1989), Nei's D averaged 0.063 between single samples of *P. inornatus* and *P. bicolor*. These two species of crested titmice (subgenus *Baeolophus*) formed a distinct lineage from other parids based on allozymes (ibid.).

Phenograms and Phylogenetic Relationships

The UPGMA phenogram based on Cavalli-Sforza and Edwards (1967) chord distance (Table 17) revealed three major clusters of allozymically differentiated populations (Fig. 36): A—samples from the Pacific slope region (1 to 27, Mount Dome [3] excluded) plus the Eastern Mojave (37); B—samples from the western Great Basin (31-36), Southeastern Nevada (39), and Ruby Mountains (40); and C—samples from North-central Utah (42) to South-central New Mexico (59). The two clusters (B and C) comprised of interior samples grouped together at the next level. Within each cluster, samples were mixed in terms of current subspecific classification. Thus, for example,

TABLE 17. Matrix of Nei's (1978) genetic distances (above diagonal) and Cavalli-Sforza and Edwards (1967) chord distance (below diagonal) among 32 samples of *Parus inornatus* and 2 outgroups (*P. bicolor, P. gambeli*). Sample areas are numbered as in Figs. 3-4 and Table 2.

Sample Area	1	2	3	4	5	6	10
1. Medford	-----	0.000	0.002	0.000	0.002	0.000	0.001
2. Siskiyou	0.052	-----	0.000	0.000	0.001	0.000	0.001
3. Mt. Dome	0.068	0.041	-----	0.001	0.000	0.001	0.002
4. N Sacramento Valley	0.060	0.068	0.074	-----	0.001	0.000	0.001
5. Clear Lake	0.088	0.066	0.062	0.078	-----	0.001	0.002
6. Oroville	0.066	0.077	0.078	0.051	0.091	-----	0.001
10. Monterey	0.075	0.085	0.079	0.069	0.091	0.074	-----
12. San Luis Obispo	0.063	0.069	0.084	0.056	0.087	0.071	0.057
15. W sl. Sierra Nevada-S	0.057	0.064	0.080	0.050	0.082	0.059	0.075
16. Kern	0.046	0.055	0.070	0.056	0.079	0.059	0.072
18. Tejon	0.075	0.078	0.094	0.059	0.080	0.066	0.088
24. Joshua Tree	0.052	0.051	0.075	0.059	0.075	0.075	0.068
27. San Diego-South	0.056	0.081	0.094	0.070	0.091	0.082	0.090
31. Warner	0.104	0.094	0.072	0.116	0.100	0.118	0.115
32. Tule Lake	0.092	0.078	0.046	0.087	0.077	0.090	0.085
33. Northeastern Calif.	0.125	0.105	0.073	0.120	0.096	0.122	0.112
34. Reno	0.084	0.061	0.033	0.083	0.070	0.086	0.086
35. Benton	0.086	0.071	0.042	0.090	0.076	0.094	0.091
36. White-Inyo Mts.	0.087	0.091	0.072	0.091	0.086	0.094	0.097
37. Eastern Mojave	0.061	0.048	0.054	0.069	0.073	0.075	0.082
39. Southeastern Nevada	0.104	0.100	0.076	0.104	0.092	0.107	0.098
40. Ruby Mts.	0.102	0.089	0.061	0.095	0.077	0.098	0.092
42. North-central Utah	0.088	0.091	0.075	0.086	0.089	0.094	0.086
44. La Sal	0.097	0.104	0.082	0.083	0.089	0.086	0.078
45. New Castle	0.092	0.092	0.075	0.080	0.088	0.083	0.077
48. N-central New Mexico	0.096	0.087	0.074	0.086	0.079	0.089	0.082
49. Zuni Mts.	0.083	0.094	0.083	0.080	0.079	0.088	0.079
52. Kaibab	0.085	0.093	0.078	0.082	0.097	0.080	0.078
53. San Francisco Mts.	0.106	0.099	0.082	0.088	0.083	0.096	0.084
56. Reserve	0.105	0.100	0.087	0.087	0.083	0.097	0.089
58. Chiricahua Mts.	0.075	0.068	0.051	0.073	0.077	0.077	0.077
59. S-central New Mexico	0.110	0.105	0.091	0.093	0.087	0.098	0.097
Parus bicolor	0.234	0.237	0.240	0.234	0.246	0.238	0.240
Parus gambeli	0.446	0.443	0.446	0.434	0.435	0.439	0.439

TABLE 17 (continued)

12	15	16	18	24	27	31	32	33	34
0.000	0.000	0.000	0.000	0.000	0.000	0.009	0.003	0.012	0.004
0.001	0.000	0.000	0.001	0.000	0.001	0.007	0.002	0.009	0.002
0.002	0.002	0.001	0.002	0.002	0.002	0.004	0.000	0.005	0.000
0.000	0.000	0.000	0.000	0.000	0.001	0.009	0.003	0.011	0.003
0.002	0.002	0.002	0.001	0.001	0.002	0.005	0.001	0.006	0.001
0.000	0.000	0.000	0.000	0.000	0.000	0.009	0.002	0.010	0.003
0.000	0.001	0.001	0.001	0.001	0.001	0.009	0.003	0.010	0.003
-----	0.000	0.000	0.000	0.000	0.001	0.010	0.004	0.012	0.004
0.049	-----	0.000	0.000	0.000	0.000	0.010	0.003	0.012	0.004
0.066	0.061	-----	0.000	0.000	0.000	0.010	0.003	0.012	0.004
0.063	0.049	0.063	-----	0.000	0.000	0.010	0.004	0.013	0.004
0.039	0.050	0.056	0.065	-----	0.001	0.010	0.004	0.012	0.004
0.070	0.065	0.073	0.067	0.071	-----	0.009	0.004	0.013	0.004
0.123	0.120	0.118	0.129	0.122	0.112	-----	0.002	0.001	0.001
0.100	0.096	0.094	0.108	0.099	0.108	0.066	-----	0.002	0.000
0.131	0.128	0.123	0.137	0.130	0.137	0.066	0.046	-----	0.002
0.092	0.089	0.086	0.101	0.091	0.101	0.057	0.035	0.057	-----
0.100	0.097	0.095	0.108	0.099	0.100	0.037	0.038	0.053	0.022
0.105	0.102	0.100	0.113	0.104	0.099	0.082	0.074	0.091	0.072
0.072	0.068	0.058	0.075	0.071	0.079	0.089	0.074	0.097	0.054
0.116	0.113	0.110	0.119	0.118	0.111	0.069	0.069	0.074	0.070
0.109	0.106	0.100	0.116	0.108	0.116	0.081	0.055	0.066	0.054
0.096	0.092	0.095	0.103	0.093	0.097	0.088	0.066	0.089	0.075
0.099	0.095	0.093	0.096	0.103	0.102	0.110	0.071	0.097	0.085
0.091	0.087	0.088	0.087	0.098	0.095	0.099	0.069	0.093	0.075
0.097	0.093	0.091	0.103	0.102	0.106	0.101	0.080	0.096	0.073
0.091	0.087	0.082	0.091	0.089	0.088	0.106	0.080	0.104	0.087
0.093	0.085	0.087	0.098	0.092	0.102	0.108	0.072	0.101	0.082
0.099	0.096	0.104	0.103	0.103	0.108	0.104	0.076	0.091	0.082
0.097	0.093	0.103	0.100	0.099	0.106	0.110	0.082	0.099	0.088
0.084	0.080	0.077	0.094	0.083	0.094	0.083	0.054	0.082	0.050
0.103	0.096	0.109	0.096	0.105	0.108	0.111	0.083	0.101	0.091
0.237	0.235	0.235	0.240	0.237	0.235	0.256	0.246	0.260	0.243
0.433	0.433	0.445	0.429	0.436	0.436	0.454	0.447	0.452	0.448

TABLE 17 (continued)

Sample Area	35	36	37	39	40	42	44
1. Medford	0.004	0.003	0.000	0.004	0.005	0.002	0.004
2. Siskiyou	0.002	0.002	0.000	0.003	0.003	0.002	0.003
3. Mt. Dome	0.000	0.001	0.000	0.001	0.001	0.001	0.002
4. N Sacramento Valley	0.004	0.003	0.000	0.004	0.004	0.002	0.003
5. Clear Lake	0.002	0.001	0.001	0.001	0.001	0.002	0.002
6. Oroville	0.004	0.002	0.000	0.004	0.004	0.002	0.002
10. Monterey	0.004	0.003	0.001	0.004	0.004	0.002	0.002
12. San Luis Obispo	0.005	0.004	0.001	0.005	0.005	0.002	0.004
15. W sl. Sierra Nevada-S	0.005	0.003	0.000	0.005	0.005	0.002	0.003
16. Kern	0.005	0.003	0.000	0.005	0.005	0.002	0.003
18. Tejon	0.005	0.004	0.001	0.005	0.006	0.003	0.004
24. Joshua Tree	0.005	0.004	0.001	0.005	0.005	0.002	0.004
27. San Diego-South	0.005	0.003	0.001	0.005	0.006	0.003	0.004
31. Warner	0.001	0.003	0.006	0.001	0.003	0.006	0.006
32. Tule Lake	0.000	0.000	0.001	0.000	0.000	0.000	0.000
33. Northeastern Calif.	0.001	0.004	0.008	0.002	0.002	0.005	0.005
34. Reno	0.000	0.001	0.001	0.000	0.000	0.002	0.002
35. Benton	-----	0.001	0.002	0.000	0.000	0.002	0.002
36. White-Inyo Mts.	0.067	-----	0.002	0.000	0.000	0.001	0.001
37. Eastern Mojave	0.065	0.081	-----	0.002	0.002	0.002	0.003
39. Southeastern Nevada	0.061	0.081	0.097	-----	0.000	0.001	0.001
40. Ruby Mts.	0.057	0.078	0.085	0.066	-----	0.002	0.001
42. North-central Utah	0.072	0.089	0.091	0.079	0.079	-----	0.000
44. La Sal	0.087	0.092	0.096	0.084	0.080	0.067	-----
45. New Castle	0.078	0.095	0.083	0.087	0.086	0.068	0.048
48. N-central New Mexico	0.079	0.090	0.083	0.077	0.078	0.089	0.093
49. Zuni Mts.	0.086	0.089	0.091	0.079	0.088	0.073	0.068
52. Kaibab	0.085	0.090	0.091	0.093	0.085	0.045	0.061
53. San Francisco Mts.	0.083	0.096	0.100	0.075	0.074	0.055	0.067
56. Reserve	0.090	0.097	0.102	0.082	0.089	0.073	0.080
58. Chiricahua Mts.	0.056	0.078	0.063	0.077	0.069	0.058	0.077
59. S-central New Mexico	0.092	0.105	0.104	0.099	0.096	0.070	0.072
Parus bicolor	0.246	0.248	0.236	0.252	0.249	0.246	0.248
Parus gambeli	0.449	0.447	0.447	0.445	0.448	0.441	0.437

TABLE 17 (continued)

45	48	49	52	53	56	58	59	bicolor	gambeli
0.002	0.003	0.001	0.001	0.003	0.002	0.001	0.003	0.063	0.278
0.001	0.002	0.001	0.001	0.002	0.002	0.000	0.002	0.064	0.276
0.001	0.000	0.001	0.001	0.001	0.001	0.000	0.001	0.065	0.279
0.001	0.002	0.001	0.001	0.002	0.001	0.001	0.003	0.062	0.271
0.001	0.001	0.001	0.002	0.001	0.000	0.000	0.001	0.066	0.268
0.001	0.002	0.000	0.001	0.002	0.001	0.000	0.002	0.064	0.274
0.001	0.002	0.001	0.001	0.002	0.002	0.001	0.003	0.064	0.271
0.002	0.003	0.001	0.002	0.003	0.002	0.001	0.003	0.064	0.270
0.001	0.002	0.001	0.001	0.003	0.002	0.001	0.003	0.063	0.272
0.001	0.002	0.001	0.001	0.003	0.002	0.001	0.003	0.063	0.276
0.001	0.003	0.001	0.002	0.003	0.002	0.002	0.003	0.064	0.264
0.002	0.003	0.001	0.002	0.003	0.002	0.001	0.003	0.064	0.269
0.002	0.003	0.001	0.002	0.004	0.003	0.002	0.004	0.064	0.273
0.006	0.004	0.007	0.007	0.005	0.006	0.005	0.005	0.074	0.289
0.001	0.000	0.001	0.001	0.000	0.000	0.000	0.000	0.067	0.280
0.006	0.004	0.007	0.006	0.004	0.005	0.005	0.004	0.076	0.286
0.002	0.000	0.002	0.002	0.001	0.001	0.000	0.002	0.067	0.283
0.002	0.001	0.002	0.003	0.001	0.002	0.001	0.002	0.068	0.283
0.002	0.001	0.001	0.001	0.000	0.001	0.001	0.001	0.068	0.280
0.001	0.001	0.001	0.001	0.002	0.002	0.000	0.002	0.064	0.278
0.002	0.000	0.001	0.002	0.000	0.001	0.001	0.001	0.067	0.278
0.002	0.000	0.002	0.002	0.000	0.001	0.001	0.001	0.068	0.282
0.000	0.001	0.000	0.000	0.000	0.000	0.000	0.000	0.066	0.272
0.000	0.001	0.000	0.000	0.000	0.000	0.001	0.000	0.067	0.270
-----	0.001	0.000	0.000	0.000	0.000	0.000	0.000	0.065	0.264
0.078	-----	0.001	0.001	0.000	0.000	0.000	0.001	0.065	0.275
0.077	0.091	-----	0.000	0.000	0.000	0.000	0.000	0.062	0.271
0.069	0.094	0.075	-----	0.000	0.000	0.000	0.000	0.064	0.274
0.069	0.069	0.085	0.069	-----	0.000	0.000	0.000	0.067	0.267
0.073	0.069	0.075	0.084	0.064	-----	0.000	0.000	0.063	0.265
0.064	0.065	0.070	0.063	0.073	0.076	-----	0.001	0.063	0.278
0.065	0.091	0.086	0.077	0.070	0.071	0.077	-----	0.067	0.263
0.246	0.246	0.243	0.243	0.252	0.249	0.239	0.253	-----	0.268
0.434	0.440	0.440	0.445	0.431	0.430	0.446	0.430	0.436	-----

samples of "*P. i. transpositus*" from Tejon (18) and southern San Diego (27) did not group together as one would expect based on taxonomy. Similarly, the sample of "*sequestratus*" from Medford (1) clustered with "*kernensis*" from Kern (16) rather than with "*sequestratus*" from Siskiyou (2); the latter sample clustered with "*ridgwayi*" from the Eastern Mojave (37). Birds from the White-Inyo Mountains (36) did not cluster with other "*zaleptus*" in group B. Elsewhere in the interior, individuals of "*plumbescens*" from the Chiricahua Mountains (58) allied with "*ridgwayi*" from North-central Utah (42) and Kaibab (52) instead of with "*plumbescens*" from the San Francisco Mountains (53) or Reserve (56). The Mount Dome sample (3) clustered outside of all three groups at a very large genetic distance.

Similar patterns of variation emerged from the neighbor-joining (NJ) method of phylogenetic reconstruction (Fig. 37; Saitou and Nei 1987). Although the UPGMA and NJ trees differed somewhat at the level of fine branch tips, both approaches were consistent in their definition of major groupings. Interior samples of titmice formed a distinct group in the NJ tree that was clearly divergent from Pacific slope samples. This group was comprised of two subclusters that corresponded roughly to clusters B and C in the UPGMA tree. One difference occurred in the placement of samples from North-central New Mexico (48) and the Chiricahua Mountains (58), which allied with samples from eastern California and Nevada (31-36, 39-40) rather than with those from the eastern Great Basin to the Rocky Mountains and Southwest. However, the short branch length (0.0002) casts doubt on this relationship. As in the UPGMA analysis, samples from Southeastern Nevada (39) and the Ruby Mountains (40) grouped with populations from the western Great Basin. Samples from Mt. Dome (3) and the Eastern Mojave (37) were placed at intermediate links between Pacific slope and interior populations. The long branch length (0.0778) leading to Mount Dome attests to the strong allozymic divergence of this population and supports its external position in the UPGMA tree. Although Clear Lake (5) clustered with Mount Dome in the NJ tree, the branch length was relatively short (0.0007). Birds from the Eastern Mojave were most similar to those from Siskiyou (2) based on the NJ method, which agrees with the UPGMA result. In general, populations from the Pacific slope appeared to be less divergent (i.e., shorter branch lengths) than those from the interior. Two exceptions included Monterey (10) and southern San Diego (27), which had fairly long branch lengths compared to other Pacific slope samples.

In order to test the effect of input order on each tree, the sequence of populations in the genetic distance matrix was randomized using option "J" in the NEIGHBOR program of PHYLIP (Felsenstein 1991). Several randomizations were run, and the data were reanalyzed using both the UPGMA and NJ algorithms. This procedure did not affect the tree topologies.

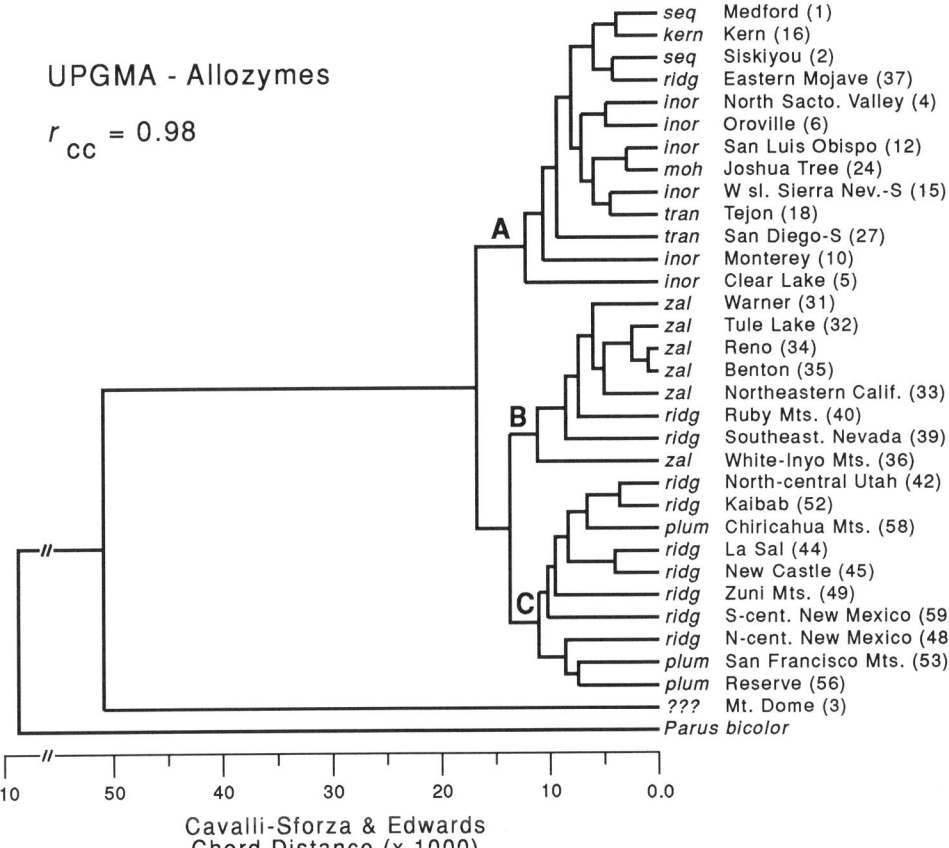

UPGMA - Allozymes

$r_{cc} = 0.98$

seq	Medford (1)
kern	Kern (16)
seq	Siskiyou (2)
ridg	Eastern Mojave (37)
inor	North Sacto. Valley (4)
inor	Oroville (6)
inor	San Luis Obispo (12)
moh	Joshua Tree (24)
inor	W sl. Sierra Nev.-S (15)
tran	Tejon (18)
tran	San Diego-S (27)
inor	Monterey (10)
inor	Clear Lake (5)
zal	Warner (31)
zal	Tule Lake (32)
zal	Reno (34)
zal	Benton (35)
zal	Northeastern Calif. (33)
ridg	Ruby Mts. (40)
ridg	Southeast. Nevada (39)
zal	White-Inyo Mts. (36)
ridg	North-central Utah (42)
ridg	Kaibab (52)
plum	Chiricahua Mts. (58)
ridg	La Sal (44)
ridg	New Castle (45)
ridg	Zuni Mts. (49)
ridg	S-cent. New Mexico (59)
ridg	N-cent. New Mexico (48)
plum	San Francisco Mts. (53)
plum	Reserve (56)
???	Mt. Dome (3)
Parus bicolor	

Cavalli-Sforza & Edwards
Chord Distance (x 1000)

FIGURE 36. UPGMA phenogram of allozyme differentiation among 32 samples of the *Parus inornatus* complex based on Cavalli-Sforza and Edwards (1967) chord distance, with *P. bicolor* as the outgroup. *Parus gambeli* was excluded from this analysis because of the relatively large genetic distances separating this species from *P. inornatus*, thus obscuring phenetic patterns within the ingroup. The high cophenetic correlation coefficient (r_{cc}) indicates excellent agreement with the original distance matrix. Abbreviations for subspecific names (American Ornithologists' Union 1957) associated with each sample are indicated. Sample areas are numbered as in Figs. 3-4 and Table 2. The Mount Dome sample (3) cannot be assigned clearly to any subspecies based on the allozyme data. Bold letters designate three major clusters of populations that are distinguishable electrophoretically.

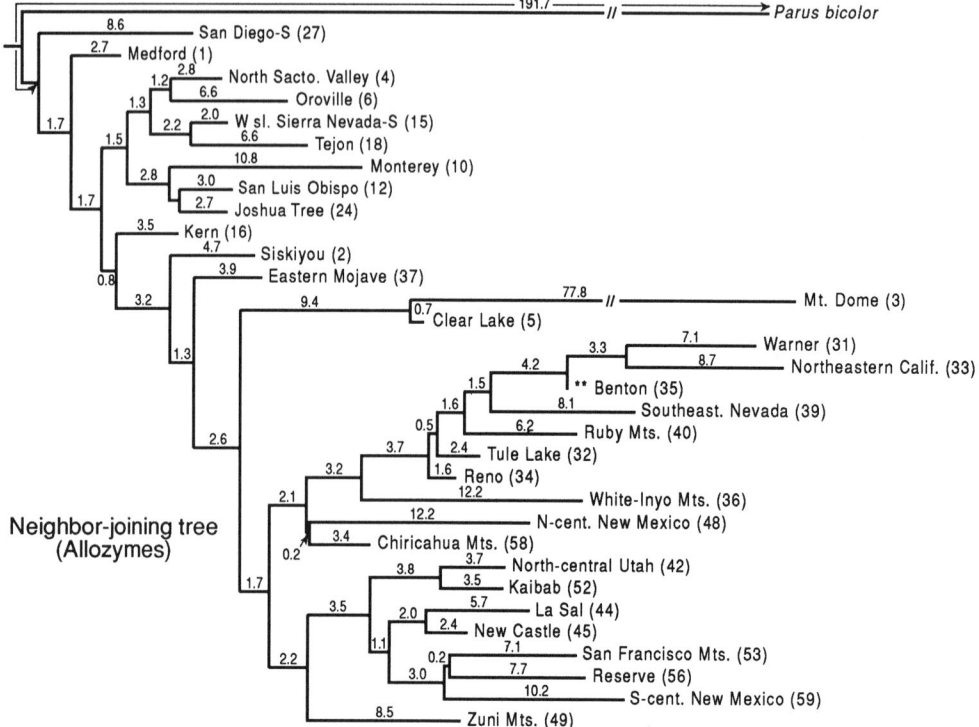

FIGURE 37. Neighbor-joining tree illustrating allozymic differentiation among 32 samples of the *Parus inornatus* complex and one outgroup (*P. bicolor*) based on Cavalli-Sforza and Edwards (1967) chord distance. Sample areas are numbered as in Figs. 3-4 and Table 2. Numbers above the branches indicate branch lengths (x 1000). Note the intermediate position and long branch length of the Mt. Dome sample (3).

Multidimensional Scaling Analysis of Genetic Distances

In a multidimensional scaling (MDS) analysis using Cavalli-Sforza and Edwards (1967) chord distances, the first axis separated Mount Dome (sample area 3; score = -7.002) from all other samples of *Parus inornatus* (scores = -0.018 to 0.780). The best resolution of allozymic variation among the 32 samples occurred along axes II and III of the MDS plot (Fig. 38). Results clearly supported the UPGMA tree in that the three major clusters of populations (A, B, C) sorted out completely in multidimensional space. The Eastern Mojave (37) sample again grouped with those from the Pacific slope (group A) rather than from the interior (B and C), and was closest to the sample from Siskiyou (2). As in the NJ tree, Clear Lake (5) stood apart from other coastal samples and showed a similarity to Mount Dome. Although Mount Dome was nearest to samples from Clear Lake (5) and the Chiricahua Mountains (58), it fell outside of the three major groupings. More importantly, Mount Dome appeared to be intermediate between groups comprised of Pacific slope and interior populations. This critical finding was implied by the neighbor-joining tree but not revealed by the UPGMA analysis. Multidimensional scaling is more useful than phenetic or phylogenetic techniques for examining non-hierarchical patterns of genetic variation, particularly if reticulate patterns are suspected (Lessa 1990).

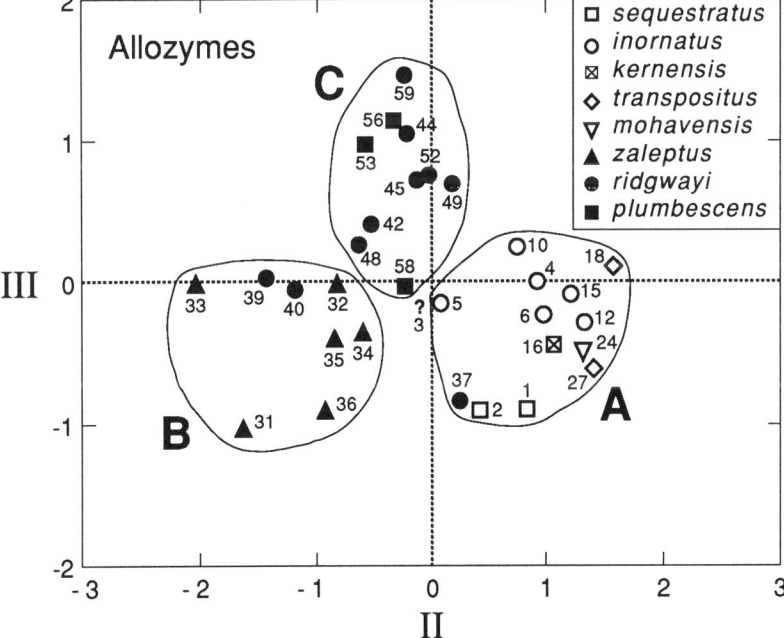

FIGURE 38. Multidimensional scaling plot of allozyme variation along axes II and III for 32 samples of *Parus inornatus*. Symbols denote current subspecies (see Figs. 21-22). Samples are numbered as in Figs. 3-4 and Table 2. Ellipses indicate major clusters identified by the UPGMA analysis (Fig. 36). The questionable genetic affinity of titmice from Mount Dome (3) is indicated.

Genetic Population Structure and Gene Flow

The 32 populations of *P. inornatus* sampled electrophoretically showed clear genetic subdivision across all polymorphic loci (F_{ST} = 0.106; Table 14). Calculation of F_{ST} for different subsets of populations revealed geographic unevenness in the amount of genetic fragmentation, however (Table 18). Overall, F_{ST} was higher in the interior (0.072) compared to the Pacific slope region (0.059). Within the interior, F_{ST} varied from 0.033 to 0.069 depending on the set of geographic comparisons; F_{ST} was higher among samples from the western Great Basin (31-37) and eastern Nevada (39-40) than from the eastern Great Basin to the Rocky Mountains and Southwest (42-59). Although the Eastern Mojave sample (37) grouped with Pacific slope samples electrophoretically, it is clearly interior on the basis of morphometrics, color, and distribution. Therefore, it was included in the F_{ST} analysis of samples from the western Great Basin to eastern and southeastern Nevada. F_{ST} changed from 0.069 to 0.053 when this sample was omitted from the subset.

In the Pacific slope region, Siskiyou (2) was more divergent from Medford (sample area 1; F_{ST} = 0.037) than from the North Sacramento Valley (4; F_{ST} = 0.028). Other Pacific slope populations from the North Sacramento Valley to southern San Diego (27) were even more fragmented (F_{ST} = 0.057). A similar value (F_{ST} = 0.052) was obtained when comparing samples from southern and southeastern California (16-27, 37). This analysis is of interest because of the phenotypic changes that occur there, both along the Pacific slope and from the coast to the interior. A comparison between Joshua Tree (24) and the Eastern Mojave (37) yielded essentially the same value (F_{ST} = 0.053). Populations from southern Oregon to northern and Northeastern California (1-3, 31-33) also showed pronounced phenotypic differences, and genetic fragmentation among these samples (F_{ST} = 0.147) was higher than for any other subset as well as for the species as a whole. However, samples across the Modoc Plateau from Siskiyou (2) to Tule Lake (32) had lower levels of divergence (F_{ST} = 0.047) compared to the entire subset. Thus, birds from Medford (1), Warner (31), and Northeastern California (33) seem to contribute substantially to estimates of genetic divergence in this region.

Gene flow estimation using S. Wright's (1951) formula ($Nm = 1/4[1/F_{ST} - 1]$) assumes an infinite island model where all loci are selectively neutral, populations are in equilibrium, and divergence is due solely to genetic drift. Although this method underestimates gene flow for populations with a one-dimensional stepping-stone structure (Slatkin 1985a), Crow and Aoki (1984) have shown that a two-dimensional stepping-stone model, which probably best approximates gene flow in the *P. inornatus* complex, produces results similar to those for the *n*-island model. Departure from the assumptions of neutrality and equilibria can influence the accuracy of gene flow estimates determined from S. Wright's (1951) F_{ST}. Nonetheless, these estimates, in conjunction with those obtained using Slatkin's (1981, 1985b) methods, enable inferences concerning past and/or present levels of gene flow among populations of titmice in western North America.

TABLE 18. F_{ST} and estimates of gene flow for different subsets[a] of 32 samples of the *Parus inornatus* complex analyzed electrophoretically.

Geographic Portion of Range	Inclusive Sample Areas	F_{ST}	Estimate of Gene Flow[b]
Entire Range	All sample areas	0.106	2.11
Pacific Slope Region[c]	1-2, 4-6, 10, 12, 15-16, 18, 24, 27	0.059	3.99
Medford to Siskiyou	1-2	0.037	6.51
Siskiyou to N Sacto. Valley	2, 4	0.028	8.68
N Sacto. Valley to San Diego	4-6, 10, 12, 15-16, 18, 24, 27	0.057	4.14
Interior Region[c]	31-37, 39-40, 42, 44-45, 48-49, 52-53, 56, 58-59	0.072	3.22
W Great Basin to E & SE Nevada	31-37, 39-40	0.069	3.37
E Great Basin to Rocky Mts., Southwest	42, 44-45, 48-49, 52-53, 56, 58-59	0.033	7.33
S Oregon to N & NE California	1-3, 31-33	0.147	1.45
Siskiyou to Tule Lake	2-3, 32	0.047	5.07
S & SE California	16, 18, 24, 27, 37	0.052	4.56
Joshua Tree to Eastern Mojave	24, 37	0.053	4.47

[a] Logical subsets were determined on the basis of electrophoretic results and barriers to gene flow. For example, samples from Siskiyou (2) and the North Sacramento Valley (4) were compared because these populations are isolated from each other by cool coniferous forest on the slopes of Mt. Shasta. Likewise, Joshua Tree (24) is separated from the Eastern Mojave (37) by the Mojave Desert.

[b] Calculated from S. Wright's (1951) formula $1/4(1/F_{ST} - 1)$.

[c] The Mt. Dome sample (3) was excluded from analyses for the Pacific slope and interior regions because it appears to receive genetic input from both directions. See text regarding the Eastern Mojave sample (37).

TABLE 19. Estimates of gene flow based on the average frequency of private alleles in 32 samples of the *Parus inornatus* complex. $Nm_c = Nm$ corrected for sample size (see Table 2 for sample sizes; average $n = 16.4$). Sample areas are numbered as in Figs. 3-4 and Table 2.

Sample Area	Single Populations			Combined Populations (one excluded)		
	No. Private Alleles	$p(1)$	Nm_c	No. Private Alleles	$p(1)$	Nm_c
1. Medford	0	----	----	15	0.042	6.360
2. Siskiyou	0	----	----	15	0.042	6.566
3. Mt. Dome	0	----	----	15	0.042	6.487
4. N Sacramento Valley	1	0.063	2.899	14	0.041	6.845
5. Clear Lake	0	----	----	16	0.041	6.727
6. Oroville	2	0.036	8.780	13	0.043	6.089
10. Monterey	0	----	----	15	0.042	6.372
12. San Luis Obispo	1	0.031	11.807	14	0.043	6.141
15. W sl. Sierra Nevada-S	1	0.033	10.432	14	0.043	6.186
16. Kern	0	----	----	15	0.042	6.360
18. Tejon	0	----	----	16	0.041	6.663
24. Joshua Tree	0	----	----	15	0.042	6.298
27. San Diego-South	2	0.044	5.901	13	0.042	6.476
31. Warner	0	----	----	15	0.042	6.373
32. Tule Lake	0	----	----	16	0.041	6.611
33. Northeastern California	0	----	----	16	0.042	6.489
34. Reno	0	----	----	15	0.042	6.335
35. Benton	0	----	----	15	0.042	6.360
36. White-Inyo Mts.	0	----	----	17	0.041	6.701
37. Eastern Mojave	0	----	----	15	0.042	6.360
39. Southeastern Nevada	0	----	----	16	0.041	6.676
40. Ruby Mts.	2	0.050	4.582	13	0.041	6.733
42. North-central Utah	0	----	----	15	0.042	6.140
44. La Sal	0	----	----	15	0.042	6.373
45. New Castle	1	0.031	11.807	14	0.043	6.140

TABLE 19 (continued)

	Single Populations			Combined Populations (one excluded)		
Sample Area	No. Private Alleles	$p(1)$	Nm_c	No. Private Alleles	$p(1)$	Nm_c
48. N-central New Mexico	0	----	----	16	0.041	6.703
49. Zuni	2	0.029	13.473	13	0.044	4.867
52. Kaibab	0	----	----	16	0.044	5.936
53. San Francisco Mts.	0	----	----	15	0.042	6.385
56. Reserve	1	0.033	10.432	15	0.042	6.360
58. Chiricahua Mts.	0	----	----	15	0.042	6.323
59. S-central New Mexico	2	0.063	2.899	13	0.039	7.409
All Samples				15	0.042	7.336

Estimates of gene flow using S. Wright's (1951) formula varied from 1.45 to 8.68 immigrants per generation for the different population subsets, equaling 2.11 overall (Table 18). This value is much lower than that obtained for all samples using Slatkin's (1985b) method of combined populations ($Nm_c = 7.336$; Table 19); Nm_c for the 32 samples ranged from 4.867 to 7.409 immigrants per generation. Slatkin's (1985b) approach estimates gene flow from the frequency of private alleles found in all populations after the successive removal of single samples. For example, removal of the Clear Lake sample (5) from this analysis yielded an estimate of 6.727 immigrants per generation, based on an average frequency of 0.041 for 16 unique alleles; although Clear Lake itself had no private alleles, one allele (MPI-a) was found at Clear Lake and in one other sample (North-central New Mexico [48]), and thus became a private allele when the Clear Lake data were excluded. For this reason, more private alleles occurred in some subsamples than in all populations combined (15 unique alleles). There was no difference in gene flow among Pacific slope samples (interior populations excluded, $Nm_c = 7.804$) and interior samples (Pacific slope populations excluded, $Nm_c = 7.770$). Similar estimates of gene flow for the two regions were obtained using S. Wright's (1951) formula based on F_{ST} (3.99 versus 3.22 immigrants per generation, respectively).

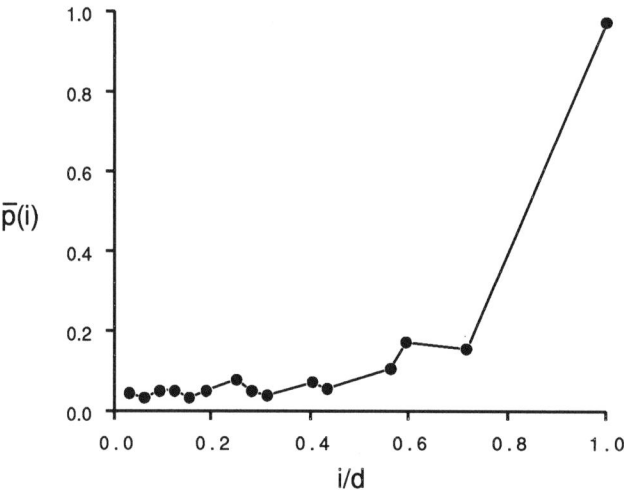

FIGURE 39. Analysis of gene flow among samples of *Parus inornatus* based on Slatkin's (1981) method. Curve depicts relationship between the average frequency of an allele ($p[i]$) and the proportion of populations (i/d) in which it occurs. Thus, alleles that are widely distributed geographically (i/d near 1.0) occur in relatively high frequency, while private or rare alleles (i/d near 0) are found in lower frequencies.

Private alleles occurred in 10 of the 32 samples at frequencies ranging from 0.029 to 0.063 (Table 19). These values translated into moderate to high estimates of gene flow for those populations (Nm_c = 2.899 to 13.473). A plot of the conditional average frequency of alleles (Fig. 39) likewise revealed high levels of gene flow (for comparison, see Figs. 1-3 in Slatkin 1981:326-329). In particular, alleles found in only one or a few samples occurred at relatively low frequency compared to alleles that had a more widespread geographic distribution. Although $p(i)$ depends strongly on migration rate (m), the influence of gene flow is dictated by the product of effective population size and migration rate (Nm) rather than by migration rate alone (Slatkin 1980, 1981).

Mantel Tests of Congruence between Allozymic and Phenotypic Traits

Pairwise comparisons of distance matrices obtained from allozyme data (Cavalli-Sforza and Edward's [1967] chord distance) and morphometric data (Euclidean distance) revealed a highly significant assocation between these suites of traits for both males (r = 0.359, t = 5.433) and females (r = 0.344, t = 5.479). In contrast, a non-significant association was found between the same genetic distance matrix and the matrix of Euclidean distances based on plumage color (r = 0.056, t = 0.555). As in the previous analyses, the *t* statistic is significant when values exceed 2.694 (*P* < 0.0071 based on the procedure-wise error rate). For the size comparisons, the probability was 1.0 that a random Mantel statistic Z is less than the observed Z; color comparisons yielded a probability of 0.711.

MITOCHONDRIAL DNA SEQUENCE VARIATION

Variation in Cytochrome *b* Sequences

Translation and alignment of sequences (53 *P. inornatus*, 1 *P. bicolor*, 2 *P. gambeli*) identified 8 distinct haplotypes that differed from each other at one or more sites (Fig. 40). Of the 300 bp analyzed, 52 (17.3%) were variable between at least two samples of the three species, and 14 (4.7%) were variable within *P. inornatus*. Levels of sequence divergence ranged from 0.3 to 4.3% in *P. inornatus*, and from 7.8 to 12.0% between members of this complex and the two outgroups (Table 20). Divergence of *inornatus* sequences was highest between haplotypes A-C and D-E ($\bar{X}$ = 4.0%). Sequences of *P. inornatus* were more similar to *P. bicolor* ($\bar{X}$ = 8.2%) than to *P. gambeli* ($\bar{X}$ = 11.33%).

Most of the changes involved silent transitions at the third position of codons (Table 20). As expected, transversions and replacement substitutions increased with percent sequence divergence. Although the ratio of transitions to transversions (transition bias) generally decreases with taxonomic distance (N.K. Johnson and Cicero 1991; M.F. Smith and Patton 1991), this was not clearly evident from the *Parus* sequences. The average transition to transversion (TS:TV) ratio between *inornatus* haplotypes A-C and D-E was 4.9 (i.e., approximately 5 transitions for every transversion). A similar value ($\bar{X}$ = 4.6) was obtained when comparing *P. inornatus* to *P. gambeli*, while comparisons between *P. inornatus* and *P. bicolor* yielded a slightly higher ratio ($\bar{X}$ = 7.4). The more closely related populations of *P. inornatus* (haplotypes A-C) had a TS:TV ratio that varied from 1:0 to 3:0. These ratios for *Parus* are similar to those obtained for comparable pairwise comparisons within and between *Amphispiza belli* and *A. bilineata* (N.K. Johnson and Cicero 1991:606).

Nucleotide Composition of Sequences

The mtDNA of vertebrates is characterized by a low percentage of guanine relative to other nucleotides at silent sites (i.e., third positions). Furthermore, avian mtDNA appears to have an unusual bias against thymine at the same position (Kocher et al. 1989). All three species of *Parus* showed a bias against guanine and thymine at third-position sites, although the extent of bias varied slightly among species. While sequences of *P. inornatus* had the lowest percentage of guanine at third positions ($\bar{X}$ = 1.0% vs. 3.1% in *P. bicolor* and 4.2% in *P. gambeli*), *P. gambeli* was especially lacking in thymine compared to the other taxa ($\bar{X}$ = 7.3% vs. 12.5% in *P. inornatus* and 14.6% in *P. bicolor*).

Sequence alignment (codon and inferred amino-acid translation) for *Psaltriparus* samples. Dots (·) denote identity with the reference sequence (*P. gambelli* 1); clover symbols (✦) mark noted positions. Sample sizes are given in parentheses after each taxon.

Block 1 (codons 35–50)

```
                      35                  40                  45                  50
                 F   G   S   L   L   G   I   C   L   I   T   Q   V   T   G   L   L   L   A
P. gambelli 1 (1)TTT GGA TCA CTC CTG GGC ATC TGC CTA ATC ACC CAA GTT ACA GGC CTA CTT CTA GCC
P. gambelli 2 (1). . . . . . . . . . . . . . . . . . .
P. bicolor    (1). .NNN . . . . . . . . . . . .A. . .C . .
P. inornatus-A(16)..C . . ..A . . . .A. ..A . . A.C . . . ..C . .
P. inornatus-B(3)..C . . ..A . . . .A. ..A . . A.C . . . ..C . .
P. inornatus-C(6)..C . . ..A . . . .A. ..A . . A.C . . . ..C . .
P. inornatus-D(27)..C . . ..A . . . .A. ..A . .G A.C . . . ..C . .
P. inornatus-E(1)..C . . ..A . . . .A.✦ ..A✦ . . A.C✦ . . . ..C . .
```

Block 2 (codons 55–70)

```
                  55                  60                  65                  70
              M   Y   T   A   D   T   L   A   F   S   V   A   H   T   C   R   N
P. gambelli 1 ATA TAC ACA GCA GAC ACC CTA GCC TTT TCC GTA GCC CAC ACC TGC CGA AAC
P. gambelli 2 . . . . . . . . . . . . . . . . .
P. bicolor    . .T . . . . . . ..C .A.T . . .T. .T. .T. . .T
P. inornatus-A. . . . . T. . . ..C .A. . . .T. .T. .T. . .T
P. inornatus-B. . . . . T. . . ..C .A. . . .T. .T. .T. . .T
P. inornatus-C. . . . .T . . . ..C .A. . . .T. .T. .T. . .T
P. inornatus-D. . . . .T . . . ..C .A. . . .T. .T. .T. . .T
P. inornatus-E. . . . .T . . .✦ ..C .A.✦ . . .T. .T. .T. . .T
```

Block 3 (codons 75–90)

```
                  75                  80                  85                  90
              V   Q   F   G   W   L   I   R   N   L   H   A   N   G   A   S   F   F   I
P. gambelli 1 GTC CAA TTC GGC TGA CTA ATC CGA AAC CTC CAT GCA AAC GGA GCC TCC TTC TTC ATC
P. gambelli 2 . . . . . . . . . . . . . . . . . . .
P. bicolor    . . .T . . .C .T .C . . .C .T .T . .T .T .T . .
P. inornatus-A. . . . . .C .T .T .T . .C .T .T .T . .T .T .T . .
P. inornatus-B..T . . . .C .T .T .T . .C .T .T .T . .T .T .T . .
P. inornatus-C. . . . . .C .T .T .T . .C .T .T .T . .T .T .T . .
P. inornatus-D..T . . . . .C .T .T .T . .C .T .T .T . .T .T .T . .
P. inornatus-E. . . . . .C .T .T .T . . .T .T .T . .T .T .T . .
```

```
                 C   I   Y   F   H   I   G   R   G   I   Y   Y   G   S   Y   L   N   K   E   T
                        95                 100                105                110
P. gambeli  1   TGC ATC TAC TTC CAC ATC GGC CGA GGG ATC TAC TAC GGC TCA TAC CTG AAC AAA GAG ACC
P. gambeli  2    .   .   .   .   .   .   .   o   .   .   .   .   o   .   .   .   .   .   .   .
P. bicolor       .   . T  .   . T . T  .   o   . A  .   .   . T  .   .   . T.A  .   .   .A.  .
P. inornatus - A .   .   . T . T  .   . T  .   .   . A . T  .   .   .   . T.A  .   .   .A.  .
P. inornatus - B .   .   .   . T  .   . T  .   .   . A  .   .   .   .   .   . T.A  .   .   .A.  .
P. inornatus - C .   .   .   .   .   . T  .   .   . A  .   .   .   .   .   .   .A.  .   .   .A.  .
P. inornatus - D .   .   .   .   .   . T  .   .   . A  .   .   . T  .   .   .   .A.  .   .   .A.  .
P. inornatus - E .   .   .   .   .   . T  .   .   . A  .   .   .   .   .   .   .A.  .   .   .A.  .


                 W   N   I   G   V   V   L   L   A   L   M   Y   A   T   A   F   V   Y   G   V
                        115                120                125                130
P. gambeli  1   TGA AAC ATT GGA GTA GTA CTT CTA GCC CTC ATA TAC GCA ACT GCC TTC GTA TAC GGA GTC
P. gambeli  2    .   .   .   .   .   .   .   .   .   .   .   .   .   .   .   .   .   .   .   .
P. bicolor       .   . T  .   .   . T .C A.  .C  . G A.   .   .   . T .C .C T  .   . G  .   .
P. inornatus - A .   . T .C  .   . C A.C .C  . A A.   .   .   .   . C .C  .   . T  .   . T  .   .
P. inornatus - B .   . T  .   .   . C A.C .C  .   A.   .   .   .   . C .C  .   . T  .   . T  .   .
P. inornatus - C .   . T  .   .   . C A.C .C  .   A.   .   .   . T . C .C  .   . C  .   . T  .   .
P. inornatus - D .   . T  .   .   . C A.C .C  .   A.   .   .   .   . C .C  .   . T  .   . T  .   .
P. inornatus - E .   . T  .   .   . C A.C .C  .   A.   .   .   .   . C .C  .   . T  .   . T  .   .
                                      ❖                 ❖
```

FIGURE 40. mtDNA sequence variation in a 300 bp fragment of the cytochrome *b* gene for 53 samples of *Parus inornatus* and 3 samples of two outgroup species, *P. bicolor* and *P. gambeli*. A total of 8 haplotypes were identified (*P. gambeli*, 2; *P. bicolor*, 1; *P. inornatus*, 5). The number of specimens represented by each haplotype is shown in parentheses. Dots indicate sequence identity to *P. gambeli*. One-letter abbreviations are given for the corresponding amino acid sequence. Symbols below blocks of sequence indicate codons with non-synonymous substitutions. Codons are numbered according to the cytochrome *b* sequence in chicken (*Gallus gallus domesticus*, Desjardins and Morais 1990).

TABLE 20. Matrix of pairwise differences and substitutions between 5 mtDNA sequence haplotypes of the *Parus inornatus* complex (A through E), 1 haplotype of *P. bicolor*, and 2 haplotypes of *P. gambeli*. Percentage sequence difference is given above the diagonal, with the ratio of transitions:transversions in parentheses. Tamura-Nei (1993) distances are given below the diagonal, with the ratio of silent:replacement substitutions in parentheses. Data are for a 300 bp fragment of the cytochrome *b* gene (see Fig. 40).

	Parus inornatus					*Parus bicolor*	*Parus gambeli 1*	*Parus gambeli 2*
	A	B	C	D	E			
inornatus - A	----	0.003 (1:0)	0.007 (2:0)	0.037 (9:2)	0.040 (10:2)	0.082 (20:4)	0.120 (29:7)	0.117 (28:7)
inornatus - B	0.003 (1:0)	----	0.010 (3:0)	0.040 (10:2)	0.043 (11:2)	0.085 (21:4)	0.117 (28:7)	0.113 (27:7)
inornatus - C	0.007 (1:1)	0.010 (2:1)	----	0.037 (9:2)	0.040 (10:2)	0.082 (20:4)	0.113 (27:7)	0.110 (26:7)
inornatus - D	0.039 (10:1)	0.042 (11:1)	0.038 (9:2)	----	0.003 (1:0)	0.078 (21:2)	0.110 (28:5)	0.107 (27:5)
inornatus - E	0.042 (11:1)	0.046 (12:1)	0.042 (10:2)	0.003 (1:0)	----	0.082 (22:2)	0.113 (29:5)	0.110 (28:5)
bicolor	0.089 (22:2)	0.094 (23:2)	0.089 (21:3)	0.086 (22:1)	0.090 (23:1)	----	0.112 (28:5)	0.109 (27:5)
gambeli 1	0.136 (30:6)	0.131 (29:6)	0.127 (29:5)	0.123 (28:5)	0.127 (29:5)	0.126 (29:4)	----	0.003 (1:0)
gambeli 2	0.131 (29:6)	0.127 (28:6)	0.123 (28:5)	0.119 (27:5)	0.123 (28:5)	0.122 (28:4)	0.003 (1:0)	----

First-position sites were comprised of approximately equal proportions of the four nucleotides ($\overline{X}$ = 25.0%), while second-position sites contained an excess of thymine relative to other bases ($\overline{X}$ = 39.6%). These data corroborate earlier studies of avian mtDNA variation (N.K. Johnson and Cicero 1991:Fig. 2; Edwards et al. 1991:Fig. 3).

Phylogeographic Patterns of mtDNA Variation

The geographic distribution of mtDNA haplotypes in *P. inornatus* revealed strong geographic structuring (Fig. 41). Of the 24 sample areas from which two or more individuals were sequenced, 21 (87.5%) showed no intra-population variability.

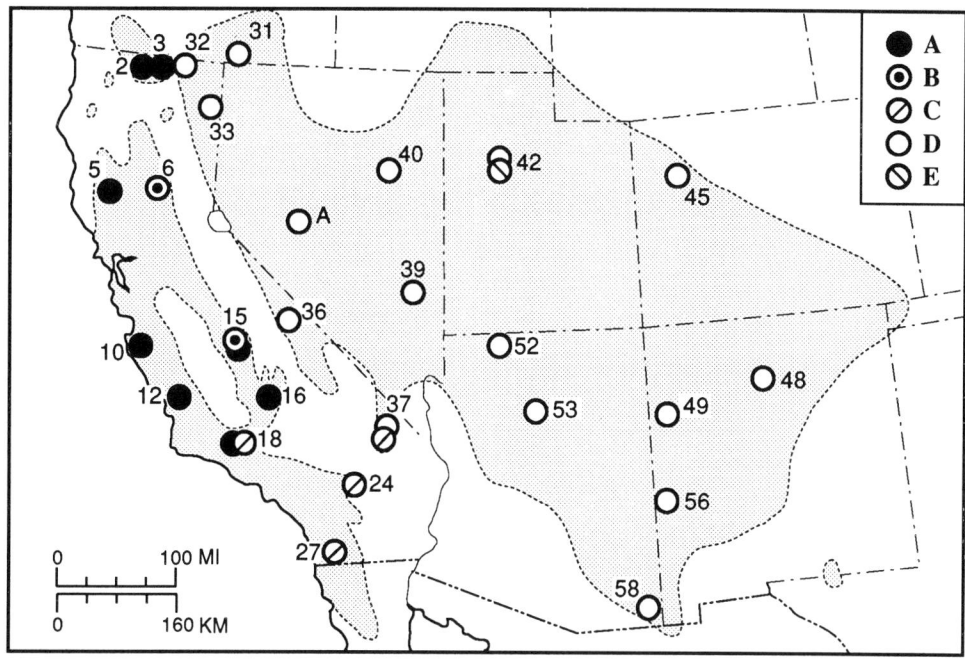

FIGURE 41. Distribution of 5 mtDNA sequence haplotypes (A-E) in 27 samples of the *Parus inornatus* complex. Sample areas are numbered as in Figs. 3-4 and Table 2; A = Fairview Peak, Churchill Co., Nevada. Note the presence of both coastal (C) and interior (D) haplotypes in the Eastern Mojave (37), and the abrupt contact of haplotypes A and D in the Modoc Plateau region of northern California (sample areas 3 [Mt. Dome] and 32 [Tule Lake], respectively).

Importantly, the three polymorphic samples were located adjacent to samples monomorphic for each type, indicating non-randomness with regard to phylogeographic pattern. Two haplotypes (A and B) were confined to samples from the Pacific slope. The most widespread of these types (A) occurred from northern California southward through the Coast Ranges, around the southern end of the San Joaquin Valley, and northward again into the southern Sierra Nevada. Both individuals from Mt. Dome (sample area 3) shared haplotype A with birds from Siskiyou (2). Haplotype B was found only in samples from the western slope of the Sierra Nevada (6, 15). A third type (C) distinguished coastal samples in southern California (including Joshua Tree [24]) from those farther north. In addition, haplotype C was observed in one of two specimens sequenced from across the desert gap in the Eastern Mojave (37). The second individual sequenced from this area shared haplotype D with all other interior samples. Another interior haplotype (E) occurred in only one specimen from North-central Utah (42).

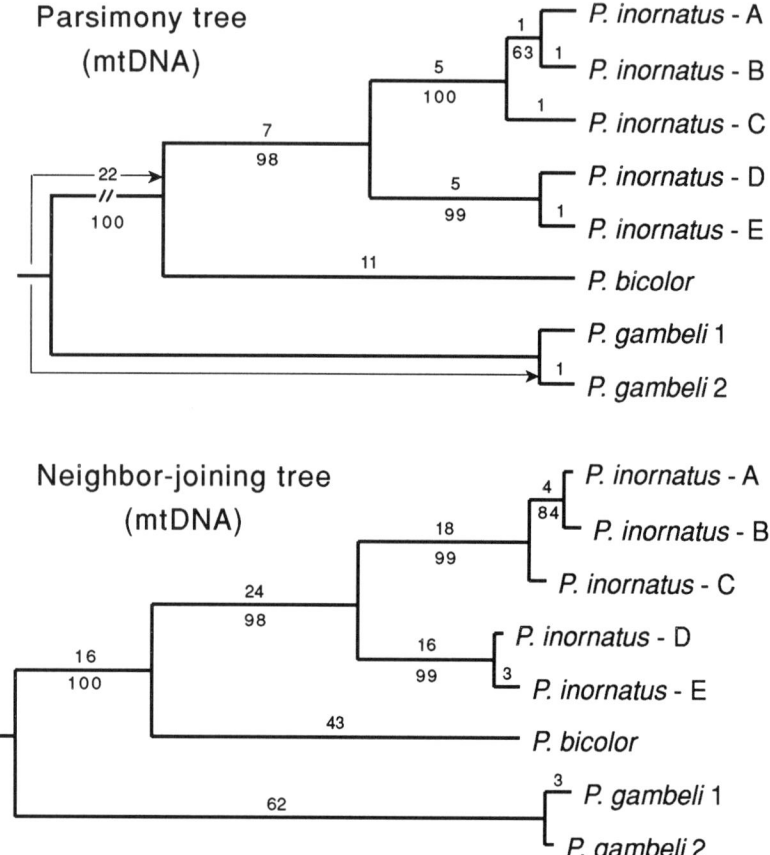

FIGURE 42. Maximum parsimony and neighbor-joining trees indicating relationships among 5 mtDNA haplotypes of *Parus inornatus*, 1 haplotype of *P. bicolor*, and 2 haplotypes of *P. gambeli* (see Figs. 40 and 41). A single tree of 55 steps was retained in the parsimony analysis (consistency index = 0.945). Numbers above the branches indicate character state changes (parsimony tree) or branch lengths (neighbor-joining tree). Numbers below the branches indicate bootstrap values for 1000 replications.

Phylogenetic Analyses

A frequency distribution of 1000 trees sampled randomly from the mtDNA sequence data was significantly non-random with regard to tree length. Likewise, an exhaustive search for the most parsimonious tree resulted in 10,395 trees (55-107 steps) with a similar pattern of distribution; only one tree was retained in this analysis. Both sets of trees were strongly skewed to the left, indicating few possible solutions for the optimal, i.e., shortest, tree (Hillis 1991; Huelsenbeck 1991; Hillis and Huelsenbeck 1992). Values of the g_1 statistic were -1.831 and -1.822, respectively, which are both highly significant at $P < 0.01$ (see Hillis 1991:Table 13-1, and Hillis and Huelsenbeck 1992:Table 2, for critical values of g_1 given a few taxa [= haplotypes, e.g., 8 in this study]). These results indicate that the *Parus* sequences contain phylogenetically informative data that can be used to infer evolutionary relationships.

Parsimony and neighbor-joining analyses identified two major clades of populations within *Parus inornatus* based on the mtDNA data (Fig. 42): haplotypes A-C, which generally characterize Pacific slope populations; and haplotypes D-E, which mark birds from the interior. As noted previously, the Eastern Mojave population (sample area 37) is polymorphic for haplotypes C and D. The strong divergence between these clades is noted by the number of changes that have occurred on the branches leading to each group. Bootstrapping with 1000 replications provided strong support for the separation of coastal versus interior clades of titmice.

The monophylly of *P. inornatus* relative to the two outgroups was supported by 98% of the trees. In concordance with the allozyme data, *P. bicolor* forms a sister taxon to all *P. inornatus* based on these mtDNA sequences. Furthermore, interior populations of *inornatus* are genetically closer to *P. bicolor* than are Pacific slope populations. *Parus gambeli* is strongly divergent from both species of crested titmice. The inclusion of one (*P. gambeli*) or both (*P. gambeli*, *P. bicolor*) outgroups did not alter the analysis.

Mantel Tests of Congruence between Allozymic and mtDNA Distances

Pairwise comparisons of distance matrices obtained from allozyme data (Cavalli-Sforza and Edward's [1967] chord distance, Nei's [1978] distance) and mtDNA sequence data (Tamura-Nei [1993] distance) revealed a highly significant assocation between patterns of genetic variation analyzed using these two approaches: chord distance, $r = 0.452$, $t = 8.076$; Nei's D, $r = 0.315$, $t = 5.554$. As in previous analyses, the t statistic is significant when values exceed 2.694 ($P < 0.0071$ based on the procedure-wise error rate). In both of these comparisons, the probability was 1.0 that a random Mantel test statistic Z is less than the observed Z.

VOCAL VARIATION

The vocalizations of *Parus inornatus* fall into two major categories (see Dixon 1949:114-116): songs, which are given by males primarily during the spring in the context of advertising or territorial defense; and call notes, which are uttered by both sexes year-round. The most frequently heard call consists of a scratchy "tsicka dee dee," which is used to defend territories, scold against intruders, and maintain contact with mates while foraging. Both sorts of vocalizations appear to be highly variable within and among individuals in a given population, although previous studies (Dixon 1969; Gaddis 1983; L.S. Johnson 1987) have focused primarily on the variety and function of male songs. Dixon (1969) recorded 17 distinct song types in a sample of about 12 males at the Hastings Natural History Reservation, Monterey County, California; individuals averaged 10 song motifs, with males typically giving up to 10 bursts of a certain song before changing to another type. Many songs were shared with birds on neighboring territories, and males often switched songs to match that of a neighbor during bouts of countersinging, resulting in a pattern of "local dialects" (Dixon 1969:99). Similar behavior was observed in a population of *P. inornatus* near Flagstaff, Arizona, where males sang an average of 13-14 song types (L.S. Johnson 1987). There, as in southeastern Arizona (Gaddis 1983), different song types were given non-randomly in association with specific behaviors.

Because these earlier studies focused on song variation within single populations of *P. inornatus*, questions pertaining to geographic variation remain largely unanswered. Thus, I initiated a preliminary study of song variation within and among populations of titmice to assess whether songs vary geographically in a pattern consistent with phenotypic and genetic traits. This relationship is of extreme interest because of the importance of song in species recognition and reproductive isolation (Payne 1986). A more comprehensive analysis of vocal variation will be presented in another paper.

A total of 2598 songs was recorded from 34 males from two coastal (8, 18) and three interior (33, 36, 56) sample areas (Table 21). Variability was higher along the Pacific slope, where 29 and 17 unique song types were observed in the San Francisco Bay region

TABLE 21. Sample sizes and song types recorded from 5 populations of the *Parus inornatus* complex.

Sample Area[a]	No. Males Recorded	No. Songs Recorded	No. Unique Song Types Recorded per Sample	Mean No. Song Types Recorded per Male[b]	Song Types[c]
San Francisco Bay Area-South (8)	11	1505	29	4.8 ± 2.9 (1-9)	A1(3), A2, A3, B1(2), C, D1, D2, D3, E1(3), E2, E3, F, G, H(2), I, J, K, L, M, N, O1(3), O2, P1, P2, P3, Q(3), R1, R2, S1, S2, T1, T2, U, V, W, X, Y, Z(2), AA, BB, CC1, CC2
Tejon (18)	10	550	17	3.2 ± 3.5 (1-12)	DD, EE1, EE2(2), FF(2), GG(2), HH, II, JJ1, JJ2, KK1, KK2, LL1, LL2(2), LL3, MM1, MM2, NN1, NN2, OO, PP1, PP2(2), QQ, RR1, RR2, SS, TT1, TT2
Northeastern California (33)	5	109	6	1.6 ± 0.5 (1-2)	UU, VV, WW4(2), XX2, YY, ZZ2(2)
White-Inyo Mts. (36)	7	431	13	2.4 ± 1.4 (1-5)	WW1, WW2, WW3, XX1(2), ZZ1, AB, AC1, AC2, AD, AE, AF, AG, AH, AI, AJ, AK
Reserve (56)	1	3	1	----	WW5
Total	34	2598			

[a] Dates of recordings are as follows: San Francisco Bay Area-South (8): 8-22 April and 6 May 1989; Tejon (18): 5-6 April 1989; Northeastern California (33): 15 June 1993; White-Inyo Mountains (36): 20 May 1991, 29-31 May 1993; Reserve (56): 6 June 1990. Ned K. Johnson assisted with recordings from Northeastern California, the White-Inyo Mountains, and Reserve.

[b] Values indicate mean $\pm$ standard deviation; range is given in parentheses. $r = 0.713$ ($P < 0.01$) between the number of songs and unique song types recorded per male.

[c] Letters designate unique song types. Minor variations on a given song type are indicated by numbers. Values in parentheses denote the number of individuals within each sample that shared a particular song. Underlined song types were shared between samples.

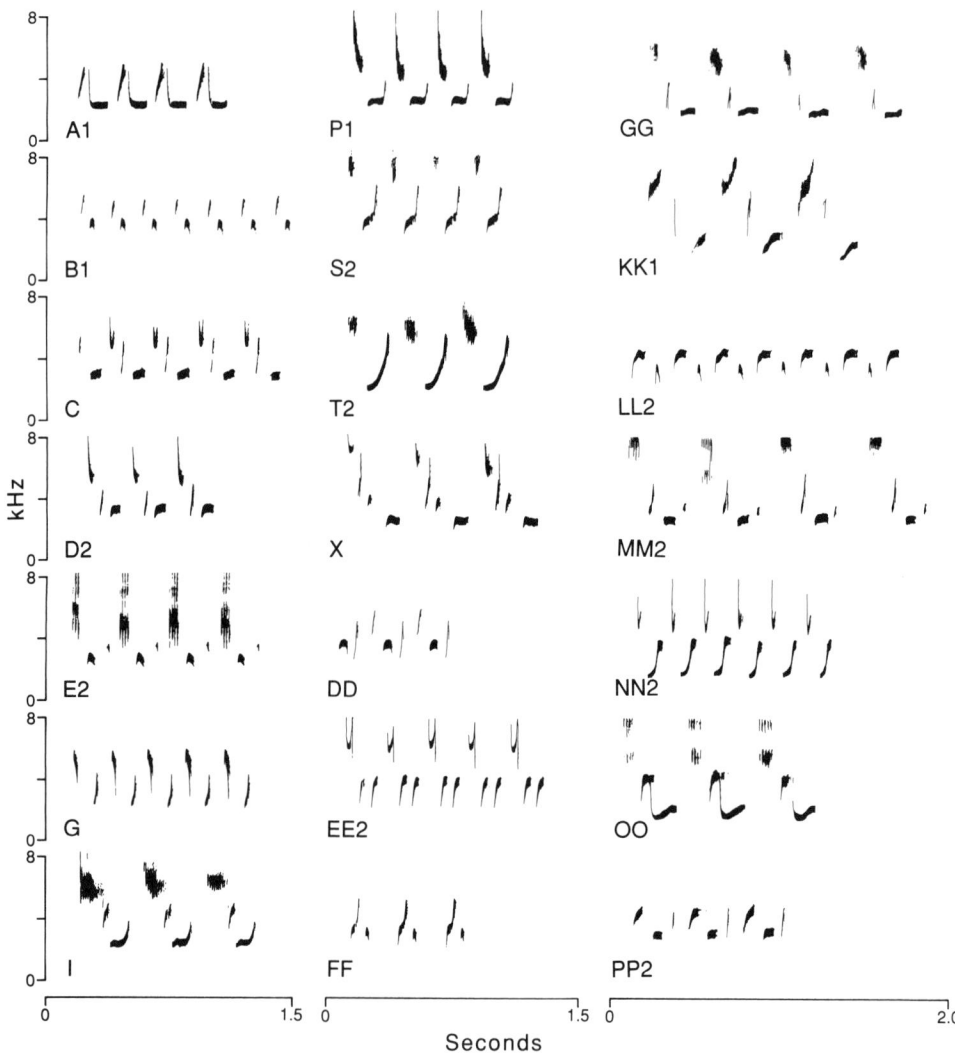

FIGURE 43. Spectrograms of songs recorded from males from two populations along the Pacific slope of California (see Table 21): San Francisco Bay Area-South [8], song types A1, B1, C, D2, E2, G, I, P1, S2, T2, X; and Tejon (18), song types DD, EE2, FF, GG, KK1, LL2, MM2, NN2, OO, PP2. Illustrated spectrograms represent the diversity of song types recorded from those areas.

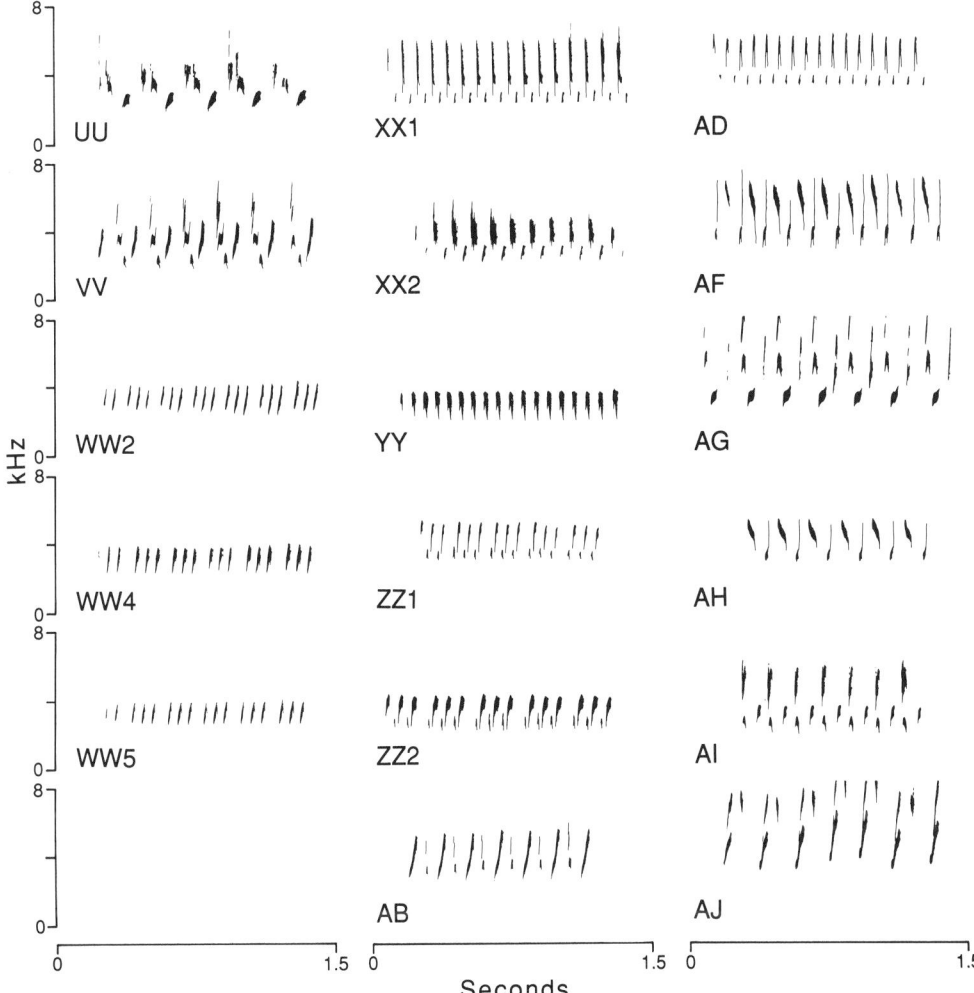

FIGURE 44. Spectrograms of songs recorded from males from two populations in the western Great Basin (Northeastern California [33], song types UU, VV, WW4, XX2, YY, ZZ2; White-Inyo Mountains [36], song types WW2, XX1, ZZ1, AB, AD, AF, AG, AH, AI, AJ) and one population in New Mexico (Reserve [36], song type WW5); see Table 21. Illustrated spectrograms represent the diversity of song types recorded from those areas. Note the similarity in structure and tempo of song types shared by different populations (e.g., types WW, XX, ZZ), and the differences in structure and tempo between these interior songs and those from the Pacific slope (Fig. 43).

and Tejon, respectively; individual males averaged 3.2 to 4.8 song types (range = 1-12). While males from Northeastern California and the White-Inyo Mountains averaged only 1.6-2.4 unique song types, smaller sample sizes may account partially for this reduced variability. Nonetheless, levels of individual variation found in this preliminary study are substantially lower than those reported by Dixon (1969) and L.S. Johnson (1987).

In both coastal and interior regions, several song types were shared by two or more males from the same population. These males typically occurred on nearby territories and, as noted by Dixon (1969) and L.S. Johnson (1987), counter-matching of songs was common. Despite the relatively large number of unique song types recorded from Pacific slope samples, no song types were shared *between* those samples. In contrast, three song types (WW, XX, ZZ) were shared by males from geographically distant populations in the interior. The occurrence of song type WW in two samples from the western Great Basin and in one sample from New Mexico is especially noteworthy.

Each song type is composed of a distinct syllable or note-complex that is produced in repeated sequence with a definite meter. Differences among songs include variability in one or more of the following features: (a) the structure of the syllables comprising the song; (b) the mean frequency (kiloHertz) of each syllable and the span of frequencies covered; (c) the number of repetitions of the same syllable; (d) the interval between syllables or sets of syllables; and (e) the duration of repetitive sequences. Preliminary results suggest that coastal and interior titmice differ in the fundamental structure and tempo of their song repertoires. Songs given by males from the Pacific slope (Fig. 43) consist of several syllables repeated at regularly spaced intervals, with different numbers of syllables distinguishing songs of one type. In general, these syllables are characterized by alternating notes of high and low frequency that sound like a ringing *pee-two pee-two pee-two* or *weety weety weety*. Songs of interior titmice (Fig. 44), on the other hand, typically lack the clear separation between high and low frequency notes. Rather, such songs sound more like a rattle or roll that consists of syllables given in rapid succession over a steady frequency range. Although males from the interior occasionally gave songs with a coastal structure (e.g., song type UU), the characteristic rattle of interior songs was not recorded in any samples from the Pacific slope.

Despite the small sample sizes and incomplete geographic representation included in the analysis, these preliminary data suggest that fundamental differences occur in the song repertoires of coastal versus interior titmice. Song types given by titmice from the two regions are presumably homologous in that they are given solely by males in the context of advertising or territorial defense during the breeding season. In light of the strong genetic and phenotypic differences that also distinguish titmice from the interior and Pacific slope, song could play an important role in reproductive isolation.

DISCUSSION AND CONCLUSIONS

Two primary goals of studies of geographic variation are to document patterns of differentiation among populations and to use such patterns to infer historical processes. Early workers were motivated largely by taxonomic and/or biogeographic interests, and the resulting studies were mainly descriptive in nature. These studies led to a plethora of new subspecies named on the basis of often subtle differences in external morphology, and to a substantial body of information concerning the distributions of species and subspecies. Within the past quarter-century, major methodological advances have revolutionized the study of geographic variation. In particular, the use of computers to analyze character variation in multivariate space, and the application of biochemical techniques to evolutionary and systematic problems, have enabled students of geographic variation to ask a variety of questions that could not be examined previously (for reviews, see Gould and Johnston 1972; Zink and Remsen 1986; Zink 1989; Avise 1994). Many of these studies addressed the importance of gene flow, or lack thereof, in promoting and/or maintaining geographic differences among populations (Slatkin 1985a, 1987, 1989; Rockwell and Barrowclough 1987). Likewise, numerous investigations focused on understanding the relative contribution of genetic versus non-genetic components of phenotypic variation (James 1970, 1982, 1983; Via and Lande 1985; Zink 1986, 1989). Regardless of the underlying questions, a clear outcome of recent research is the recognition that different characters often show discordant patterns of variation within and between species (Hillis 1987; Cheverud 1988; Shaffer et al. 1991). For this reason, studies based on several character suites provide a stronger framework for inferring systematic relationships and modes of divergence. Sets of phenotypic and genetic traits that vary concordantly among populations are especially informative in this regard.

PHENOTYPIC VARIATION

Various hypotheses have been proposed to explain geographic variation in the size and/or color of birds. While most of these have invoked extrinsic factors such as climate (e.g., Johnston 1969; James 1970; Power 1970; Aldrich and James 1991), food resources (e.g., Moen 1991), or habitat (e.g., Gibson and Kessel 1989; J. Martin 1991) to account for the observed patterns, others have explained the geographic differences in terms of intrinsic factors such as migratory distance (e.g., Wiedenfeld 1991). Traditionally, workers have assumed that phenotypic variants reflect genetically based adaptations to their environment. Support for this viewpoint comes from studies that have demonstrated heritability of phenotypic traits within populations (e.g., J. Smith and Dhondt 1980; Dhondt 1982; Davies et al. 1988), or that have shown strong correlations between morphologic and genetic patterns of variation among populations (e.g., N.K. Johnson and Marten 1992). Failure to find such evidence has motivated researchers to question this classic dogma and to search for other mechanisms of phenotypic differentiation. Thus, a number of workers have proposed that environmental induction (James 1983), phenotypic plasticity (West-Eberhard 1989), developmental shifts (Zink 1986), or random drift (A.J. Baker et al. 1990; A.J. Baker 1992) may be just as important in influencing phenotypic variability. As data on intraspecific variation in birds continue to mount, it will become increasingly apparent that numerous factors (genetic and non-genetic) can play a role and that each species must be evaluated in its own light.

Univariate patterns of size variation in titmice fell into three major categories: (1) clinal variation, which characterizes populations along the Pacific slope from the North Sacramento Valley (4) to northern Baja California (28-29); (2) disjunct or step-clinal variation, which separates Pacific slope from interior populations; and (3) patchy or mosaic variation, which generally characterizes populations from the Great Basin to the Rocky Mountains and Southwest (31-60). The smallest birds of the Pacific slope region were found on the west side of the Sierra Nevada, while the largest individuals occurred in southern California and northern Baja California. Titmice from these southern populations were noticeably larger in several characters (i.e., tail length, bill depth, bill width) than birds from other coastal populations to the north. Furthermore, they also differed significantly from the relatively small birds found in southern Baja California. Although size variation among populations showed a greater mosaic of patterns in the interior, there was a clear tendency for titmice to be larger in the more northerly sample areas (e.g., Warner, Northeastern California, Ruby Mountains, Southern Idaho, North-central Utah, New Castle). Southwestern birds overlapped in size with individuals from southern California and northern Baja California, except that the latter were distinguished by substantially longer toes. Patterns of color variation were similar to those for size, especially in terms of dorsal brightness and purity. With the exception of the Little San Bernardino Mountains, titmice from southern California and northern Baja California generally were darker than birds from other Pacific slope populations (including southern Baja California). Interior titmice were significantly brighter in coloration than their coastal relatives.

Genetically based adaptation to different environmental regimes probably accounts for much of the observed phenotypic variation between coastal and interior populations.

Importantly, numerous studies of Parids have illustrated that bill and body-size traits are heritable within populations (e.g., Garnett 1981; Dhondt 1982; Van Noordwijk 1987), and there is no reason to expect *Parus inornatus* to be an exception. Furthermore, the significant correlation (see Mantel tests) between morphometric and allozymic traits suggests that size differences among populations has at least some phylogenetic component. Critical evidence for the importance of environment in influencing phenotypic variation in titmice comes from the strong association between environmental variables and both morphometric and colorimetric traits. Titmice in the interior experience greater seasonality and more extreme winter temperatures compared to birds from the Pacific slope, factors that would influence metabolic demands (Salt 1952) and favor larger phenotypes (Murphy 1985). This is particularly true for the more northerly interior populations where, as predicted by Bergmann's ecogeographic "rule" (see James 1970), the largest individuals were observed. According to Gloger's "rule" (see Bowers 1960), the grayer plumage of interior titmice, as well as those from the Little San Bernardino Mountains, is consistent with an adaptive response to the more arid environments encountered there. Furthermore, such coloration provides better concealment against the grayish backdrop of pinyon-juniper woodland mixed with sagebrush. Populations along the Pacific slope, in contrast, generally seem to contradict Bergmann's rule. The increased size of titmice in southern California and northern Baja California may have both a phylogenetic and environmental explanation. While these populations might be derived from interior stock (see below), differences in habitat use also may be important. Block (1990) documented significant geographic variation in the ecology of *Parus inornatus* in oak woodlands of California, both in tree-species utilization and in foraging and perching substrates. Similar variability occurs in the Southwest where, for example, populations may or may not occupy pinyon-juniper woodland mixed with oaks (Dixon 1950; Marshall 1957; Gaddis 1987). Geographic differences in the dominant species composition of woodlands in the interior might also partially explain the patchiness of morphometric patterns observed there.

Some degree of phenotypic plasticity is expected in species that experience local ecological differences across their geographic distribution. Furthermore, seasonal changes in resources (e.g., food) may also influence phenotypic traits such as bill size (J. Davis 1954, 1961). The extent to which phenotypic variation in titmice is environmentally induced versus genetically controlled can best be assessed by cross-fostering and transplant experiments (e.g., James 1983; Alatalo and Gustafson 1988). In the absence of such direct evidence, one could examine whether the phenotypic differences among populations have a selective advantage such as increased fitness (e.g., Davies et al. 1988). Despite the lack of these data, the overall patterns of morphometric and colorimetric variation observed in this study reveal definite geographic structuring among populations. In conjunction with the allozyme and mtDNA data, these results have clear taxonomic implications for the species complex.

GENETIC VARIATION

Many studies of allozyme variation in birds have shown a lack of subdivision among conspecific populations, even within species that are well differentiated phenotypically (for reviews, see Barrowclough 1983; Barrowclough and Johnson 1988). Consequently, avian systematists have generally adopted the view that birds are undifferentiated allozymically relative to other vertebrates (Barrowclough 1983; Zink 1986; Barrowclough and Johnson 1988). This pattern has been attributed to high levels of gene flow and to moderate or large effective population sizes (Barrowclough 1980a; Rockwell and Barrowclough 1987; Barrowclough and Johnson 1988), all of which would promote panmixia. More recent studies, however, have demonstrated substantial allozymic differentiation in certain avian taxa (e.g., *Empidonax difficilis* [N.K. Johnson and Marten 1988]; *Amphispiza belli* [N.K. Johnson and Marten 1992]). Furthermore, newer molecular techniques such as restriction fragment analysis or DNA sequencing have provided data on genetic divergence in birds that appeared to be relatively invariant based on protein electrophoresis (e.g., Avise and Zink 1988; Tegelström et al. 1990; Moore et al. 1991; Zink 1991; Fletcher and Moore 1992). As more studies accumulate, workers are being forced to re-evaluate the traditional viewpoint that birds lack genetic population structure.

Genetic Variation Within Populations

Mean heterozygosity (0.025) of titmouse populations in the western United States was relatively low compared to other avian species (see Corbin 1983:Table 1; Barrowclough 1983:Table 7.1; Barrowclough and Gutiérrez 1990:Fig. 2). Associated measures of variability (percentage of polymorphic loci; mean number of alleles per locus) also were fairly low, which is not surprising given the high correlation among them. Despite the lower density and effective population size of titmice from the interior, intra-population heterozygosity generally was greater in that region ($\bar{H}_{obs}$ = 0.032) than along the Pacific slope ($\bar{H}_{obs}$ = 0.015). N.K. Johnson and Marten (1992) reported a similar finding for Great Basin versus coastal subspecies of *Amphispiza belli*. One reason for this trend may be the greater environmental (especially climatic) heterogeneity experienced by birds in the interior, a factor associated with increased genetic polymorphism (Antonovics 1971; Hedrick et al. 1976; Hedrick 1986; Nevo and Beiles 1989). Although a possible explanation is that the polymorphic loci are non-neutral and selectively advantageous (G.B. Johnson 1973; but see Kimura 1979, 1982; Barrowclough et al. 1985; Ohta 1992), genetic "hitch-hiking"—i.e., selection on linked genes—is more likely (see Hedrick et al. 1976). While variable environments may influence genetic diversity, experiments concerning their role in maintaining heterozygosity have resulted in conflicting and often inconclusive evidence (Hedrick 1986). Likewise, I failed to find a clear association between particular environmental variables and measures of allozymic variability, despite their high overall correlation.

Several populations of titmice had especially low levels of variability compared to others from the same geographic region (i.e., Medford [1], Kern [16], Eastern Mojave [37], Chiricahua Mountains [58]). In the Eastern Mojave and Chiricahua Mountains, titmice occur at low density and are highly isolated from other populations. Such conditions would make them vulnerable to population bottlenecks, a process commonly associated with reduced heterozygosities (Nei et al. 1975) or other measures of variability (Leberg 1992). Small population size may likewise account for low genetic variation in the Kern sample. In contrast to these three areas, population density is relatively high at Medford. Although the age of this population is unknown, a relatively recent founding event by a few individuals, or a population bottleneck, may explain the exceptionally low heterozygosity (0.006) observed in that sample.

Genetic Variation Among Populations

Allozyme analysis revealed that 10.6% of the total genetic variance was distributed among populations of titmice, illustrating a clear departure from panmixia. This value is relatively high compared to most allozymic studies of geographic variation in birds (Barrowclough 1983; Barrowclough and Johnson 1988; Barrowclough and Gutiérrez 1990:Fig. 3). However, similar or higher values have recently been reported for several other avian taxa in western North America (N.K. Johnson and Marten 1988, 1992; Barrowclough and Gutiérrez 1990; Cicero and Johnson 1992).

The most divergent populations (F_{ST} = 0.147) were those from the Pacific slope and interior regions of southern Oregon and northern California, i.e., Medford (1) and Siskiyou (2) across the Modoc Plateau to Warner (31) and Northeastern California (33). Surprisingly, a comparison of samples from southern and southeastern California (Joshua Tree [24] versus Eastern Mojave [37]) revealed only moderate allozymic differentiation (F_{ST} = 0.053), despite the extensive desert habitat separating those populations. A similar F_{ST} value (0.059) was observed when comparing all samples along the Pacific slope. The higher F_{ST} (0.072) for interior samples suggests greater allozymic fragmentation in that region, a finding that is consistent with the more patchy distribution of titmice there.

Additional evidence for the strong genetic divergence of Pacific slope and interior populations of titmice comes from mtDNA data. The 4.0% average difference in cytochrome *b* sequences between titmice from these geographic regions is substantially higher than that reported for most other conspecific populations of birds (e.g., *Amphispiza belli*, 0.6% [N.K. Johnson and Cicero 1991]; *Uria lomvia*, 1.8% [Birt-Friesen et al. 1992]). Restriction fragment analysis likewise showed high nucleotide divergence (p = 0.05) between coastal and interior samples of titmice (Gill and Slikas 1992). Interestingly, sequence divergence between these two clades is similar to that found between geographically segregated lineages of *Pomatostomas temporalis* that further exhibit an mtDNA length polymorphism (3.2% average divergence, Edwards and Wilson 1990). A comparison of cytochrome *b* sequence divergence among numerous avian congeners revealed levels ranging from 0.4% (*Sphyrapicus*) to 14.2% (*Phylloscopus*), with means of 3.3-12.9% (Cicero and Johnson 1995).

Except for the Eastern Mojave (31) and North-central Utah (42) sample areas, all titmice sequenced from the interior were identified by a single cytochrome *b* haplotype. Pacific slope populations, on the other hand, showed greater subdivision in their sequences, a finding that contradicts the results of the allozyme analysis. Gill and Slikas (1992) similarly found no variant restriction haplotypes in a sample of the interior subspecies *Parus inornatus ridgwayi*, but observed two types in a sample of coastal *P. i. transpositus*. The geographic distribution of mtDNA sequence haplotypes along the Pacific slope agrees roughly with the phenotypic patterns of variation. While the smallest birds from the western slope of the Sierra Nevada were distinguished by one haplotype (B), another haplotype (C) differentiated larger-sized titmice from southern California. Haplotype A characterized medium-sized birds from other widely separate coastal populations. Although some mixing of females is evident at the boundaries between types A and both B and C, the greater mtDNA fragmentation of Pacific slope populations supports the hypothesis that coastal titmice are more strictly sedentary than their interior relatives (see discussion of territoriality in "Synopsis of Natural History," above).

Both the allozyme and mtDNA data indicate that the primary pathway for genetic divergence in titmice occurred *between* Pacific slope and interior populations rather than within each region. In general, these data sets provide different levels of resolution for examining questions of genetic variation within and between species. Contrary to the opinion of many avian systematists, protein electrophoresis is still useful for surveying molecular variability among conspecific populations. However, mtDNA techniques enable a more refined look at genetic differentiation for a limited portion of the total genotype. Because nuclear and mtDNA markers reflect different evolutionary forces (Mitton 1994), future investigations of geographic variation will continue to benefit from both approaches in their attempts to unravel patterns of inter-population genetic divergence.

Phylogenetic Relationships of Populations

Pacific slope and interior populations of titmice clearly emerged as separate clades based on allelic frequencies and mtDNA sequences. Within the interior, another split differentiating western and eastern populations was evident from the allozyme data. The peptidase leucyl-glycyl-glycine (LGG) was most informative in identifying the western group, which includes populations ranging from the east slope of the Cascade and Sierra Nevada mountains to eastern and southeastern Nevada. Unlike the mtDNA sequence data, allozymes failed to reveal a pattern of relationships among coastal populations.

Perhaps the most surprising result was the strong allozymic similarity between titmice from the Eastern Mojave (sample area 37) and the Pacific slope. Birds from the Eastern Mojave are most similar to other interior titmice in both size and color. According to the allozyme data, one might conclude that these phenotypic patterns reflect a plastic, environmentally induced change from a coastal ancestor. However, the phylogeographic distribution of mtDNA sequences yields crucial insight into the relationship of populations in southern and southeastern California. In particular, the presence of both coastal and interior haplotypes in the Eastern Mojave provides convincing evidence for phylogenetic co-ancestry, with a southern connection between Pacific slope and interior populations during the evolutionary past. The fact that the coastal haplotype found in the Eastern Mojave is diagnostic of populations in southern California, rather than representing one of the other two haplotypes present along the Pacific slope, indicates non-randomness of historical events. Evidence for genetic introgression from the interior into the Eastern Mojave also comes from a rare allele (b) at the ADA locus, which was unique to titmice sampled from that region, the White-Inyo Mountains, and Northeastern California. Because of the strictly sedentary behavior of at least coastal titmice, the Mojave Desert acts as a major barrier to dispersal between Pacific slope and interior populations in southern and southeastern California. Thus, current gene flow across this gap is highly unlikely. Present patterns of allozymic and mtDNA variation in the Eastern Mojave undoubtedly are a consequence of past genetic exchange, probably as recently as 10,000 years ago (see discussion below on Historical Processes of Differentiation).

In light of the strong genetic differentiation between Pacific slope and interior populations of *Parus inornatus*, a relevant question concerns which of these clades is more similar to *P. bicolor*, the sister group to the whole species complex (Gill et al. 1989; Sheldon et al. 1992). Restriction fragment analysis of two subspecies each of *P. bicolor* (*P. b. bicolor*, *P. b. sennetti* [*atricristatus* group]) and *P. inornatus* (*P. i. transpositus*, *P. i. ridgwayi*) revealed moderate levels of divergence between species as determined by shared restriction fragments ($p = 0.0522$ to 0.0568; Gill and Slikas 1992). Divergence estimates based on shared restriction sites were higher and showed greater variability between subspecies of the two species (from 7.1% [*P. b. bicolor* vs. *P. i. transpositus*] to 8.6% [*P. b. sennetti* vs. *P. i. ridgwayi*]). These latter values are more in line with the average percentage sequence difference observed between *P. inornatus* and *P. bicolor* (8.2%) in the current study. According to the nucleotide data, interior populations of *Parus inornatus* are more similar to *P. bicolor* than are Pacific slope forms.

CONGRUENCE OF PHENOTYPIC AND GENOTYPIC PATTERNS

Qualitative appraisal of the patterns of morphometric and allozymic variation showed broad-scale geographic congruence, with Pacific slope and interior populations of titmice forming separate groups based on both clustering and ordination techniques. This finding was statistically corroborated by the results of Mantel tests, which demonstrated a significant association between distance matrices derived from the morphologic and allozymic data sets. Although geographic patterns of allozyme variation were not statistically correlated with patterns of plumage coloration, the univariate and multivariate analyses of color clearly revealed differences between coastal and interior titmice.

Furthermore, Mantel tests showed that patterns of color variation were geographically congruent with those exhibited by morphology. Mantel tests also revealed strong concordance between allozymic and mtDNA patterns of variation. Importantly, the phenotypic and genetic distinctions between coastal and interior titmice coincide with differences in habitat and climate between the two regions.

ZONES OF ABRUPT CHARACTER CHANGE, INTERGRADATION, AND SECONDARY CONTACT

Contact zones between well-differentiated avian populations have been the subject of numerous investigations (e.g., Dixon 1955, 1990; Moore and Buchanan 1985; Moore and Koenig 1986; Grudzien et al. 1987; Fleischer et al. 1991). In addition to providing important material for the study of speciation (Hewitt 1988, 1989; Harrison and Rand 1989), such zones can play a critical role in reinforcing reproductive isolating mechanisms that have evolved in allopatry. Because contact zones occur at the periphery of the geographic ranges of parental taxa, they often are characterized by relatively sharp environmental gradients and low population densities. As a result, these areas may act as "tension zones" in which character variation within the zone is maintained by a balance between dispersal from one or both parental forms and selection against intermediates (Barton and Hewitt 1985; Arnold 1992). During bountiful years, contact zones may provide "sinks" (sensu Pulliam 1988) for individuals dispersing from source populations where productivity exceeds mortality. These individuals may either colonize new areas or may enter habitat that is already occupied. Likewise, certain populations may become extirpated during times of low population density. Such non-equilibrium dynamics of extirpation and recolonization can critically affect local genetic population structure, especially when extirpation rates are high and when the number of individuals founding a new population is small relative to the number of migrants between extant populations (Wade and McCauley 1988; McCauley 1993). The origin and composition of colonizing individuals also have an important influence on genetic patterns at the local scale (Whitlock and McCauley 1990).

In western North America, a major center of avian differentiation occurs at the interface between the Great Basin and Cascade-Sierra Nevada mountains (A.H. Miller 1941; Behle 1963; N.K. Johnson 1978). Numerous bird species in addition to titmice show striking differences in size and/or color in this region, often with accompanying breaks in habitat preference (e.g., N.K. Johnson 1978:Table 5), allozymes (e.g., N.K. Johnson and Marten 1988), mtDNA (e.g., Zink 1991), and/or song (e.g., N.K. Johnson 1980). These patterns are attributed to steep climatic and ecologic gradients that separate the two physiogeographic regions and hinder gene flow between them. Although coastal and interior forms of some species exhibit varying degrees of primary or secondary intergradation, the subspecies of others occur in disjunct allopatry (N.K. Johnson 1978).

Prior to the current study, previous workers (Grinnell and Miller 1944:308; N.K. Johnson 1978:153) thought that Pacific slope and interior forms of *P. inornatus* were strongly allopatric. My fieldwork, however, filled in some gaps in the known distribution of the species, particularly on the Modoc Plateau in northern and northeastern California. A surprising result of this work was the discovery that coastal titmice (*P. i. sequestratus*) there occupy stands of pure western juniper in addition to junipers mixed with oaks. The lack of dependence on oaks removes a potential ecological barrier to intergradation between Pacific slope and interior (*P. i. zaleptus*) titmice in that region. However, the relatively large F_{ST} value (0.147) separating populations of *sequestratus* and *zaleptus* on and around the Modoc Plateau indicates restricted gene flow. Although the allozyme analysis revealed past introgression of coastal and interior alleles near Mt. Dome (sample area 3), mtDNA sequence data provide strong evidence for coastal ancestry of that population. The sharp break in mtDNA haplotypes on the Modoc Plateau, combined with the pattern of allelic frequencies, suggests secondary contact between formerly allopatric groups of populations that have evolved independently along the Pacific slope and in the Great Basin. This putative secondary contact zone, which has a maximum width of 20-25 mi (33-42 km), appears to be of recent origin. Paleobotanical data from Lava Beds National Monument and other sites in the northwestern Great Basin indicate that western juniper first appeared in the region between 7000 and 4000 years ago (R.F. Miller and Wigand 1994). Furthermore, the distribution and abundance of juniper have fluctuated dramatically since its arrival there (ibid.). Such fluctuations would provide tenuous and intermittent conditions for contact between Pacific slope and interior titmice. The current patchiness of juniper on the Modoc Plateau, in conjunction with an east-west climatic gradient, certainly limits free interbreeding among populations. Whereas occasional dispersal, especially by unmated juveniles, may lead to the founding of new populations and subsequent genetic introgression, the low density and scattered occurrence of titmouse populations on the Modoc Plateau enhances their probability of local extirpation.

As stated previously, another strong break in gene flow occurs across the Mojave Desert in southern and southeastern California. Although allozymic and mtDNA patterns suggest fairly recent interchange between Pacific slope and interior populations in this region, sharp phenotypic differences as well as changes in habitat preference are evident. According to Garrett and Dunn (1981:269), titmice occur "irregularly in fall and winter on the western edge of the deserts [in southern California]...." Vagrants also appear rarely near the Salton Sea in the Colorado Desert. All of these individuals are of the coastal form *transpositus*. While these few records of vagrancy reveal some potential for movement, the fact that they are restricted to the western side of the deserts indicates that the harsh environmental conditions there present a formidable barrier to cross-desert dispersal. The lack of contact between these populations should result in subsequent genetic change that, in conjunction with phenotypic and ecologic differences, would enhance their reproductive isolation.

In contrast to the Modoc Plateau and Mojave Desert areas, inter-population movement within the Coast and Transverse ranges of California is not constrained by sharp environmental gradients. Furthermore, a more continuous population structure, coupled with higher densities, probably results in greater dispersal and thereby increases the probability of mixing. Grinnell and Miller (1944:308-309) reported intergradation between the nominate subspecies *P. i. inornatus* and the southern form *P. i. transpositus* in Santa Barbara and Ventura counties, noting that "good" *inornatus* occur only as far south as "Morro, San Luis Obispo County." This region coincides approximately with the step-cline in certain phenotypic traits that I observed between northern and southern populations of coastal titmice. The relative similarity in phenotype and genotype of titmice from these populations, however, indicates a weak barrier to gene flow. Likewise, populations of small-sized titmice along the west slope of the Sierra Nevada are not isolated from those to the north (i.e., North Sacramento Valley [4]) or south (i.e., Kern [16]), and intergradation at both ends is probable. The mtDNA data provide some evidence for genetic introgression where phenotypically divergent populations of titmice meet at the southern end of the Coast and Sierra Nevada ranges.

HISTORICAL PROCESSES OF DIFFERENTIATION

Students of geographic variation have used different types of patterns to infer historical processes of reproductive isolation and evolution (Bush 1975; Endler 1977; Templeton 1980). Although most avian systematists advocate allopatric speciation models, debate persists over the role of vicariance versus dispersal in the differentiation of populations. Because species exhibit unique sets of patterns, hypotheses to account for inter-population differences must be species-specific. Nonetheless, broad congruence in patterns exhibited by unrelated taxa implies similar processes of divergence. The following section reviews the evolutionary history of woodland vegetation in western North America, discusses mechanisms and timing of differentiation in *P. inornatus*, and proposes a model to explain the distribution and divergence of populations in this species complex.

Evolution of the Madro-Tertiary Geoflora

According to fossil evidence, the sclerophyllous Madro-Tertiary Geoflora first appeared in North America during the early Tertiary and became increasingly widespread during the Miocene and Pliocene (Axelrod 1958, 1973). This flora was comprised of several woodland elements from which modern communities are derived, but which contained a richer assemblage of species with wider distributions than their descendants (Axelrod 1958, 1973, 1977). Thus, constituents of the California Woodland Element (e.g., *Quercus*) mixed with fossil representatives of the Conifer Woodland Element (pinyon-juniper) and other coastal or interior species in the Great Basin and southern California (Axelrod 1958). Significant changes in the composition and distribution of vegetation types first occurred during the middle Pliocene (3-4 MYBP; Axelrod 1958,

1973), when the climate was warmer and drier than at any other time in the Tertiary (Axelrod 1948). Increasing aridity in the mid-Pliocene was followed by the development of cold, wet winters in the late Pliocene. Cooler winter temperatures, combined with reduced summer precipitation, resulted in substantial reorganization of plant communities (Axelrod 1958). These shifts were accelerated during the Pleistocene because of climatic changes accompanied by intense topographic uplifting. The result was an impoverishment of sclerophyllous plant communities, which consisted of floras adapted to the new environments. Surviving species segregated into assemblages similar to modern analogs. Several changes that occurred during the late Pliocene and early to middle Pleistocene (Axelrod 1958, 1973) are particularly relevant to the discussion of evolution in titmice: (1) retreat of California woodland species from the Great Basin; (2) appearance of shrub steppe environments over the lowlands of the Great Basin, with subsequent confinement of woodlands to the slopes of moister basin ranges; (3) spread of oak woodland-savanna from southern California and the western foothills of the Sierra Nevada to west-central California; (4) movement of oak-gray pine woodland from southern into central California; (5) uplift of the Transverse Ranges, which resulted in a sharp floral transition between central and southern California floras; and (6) isolation of the Lagunan Woodland Element in the mountains of southern Baja California.

Glacial-Interglacial Cycles and the Pleistocene-Holocene Transition

Evidence from carbon isotope ratios in southern Nevada reveals 4 glacial-interglacial transitions during the middle and late Pleistocene (Coplen et al. 1994). At least in California, each successive interglacial period was marked by increasingly severe summer drought that now characterizes its Mediterranean climate (Axelrod 1973). Alternating periods of cool-moist (pluvial) and hot-dry (Mediterranean) climate inevitably led to repeated spatial shifts in plant (Axelrod 1973) and animal populations. Such changes have been invoked to explain patterns of geographic variation and distribution in a wide array of organisms (e.g., Rand 1948; Mengel 1964; Hubbard 1973; Kat 1985; Mayr and O'Hara 1986; Sage and Wolff 1986; Riddle and Honeycutt 1990; Bermingham et al. 1992; Nores 1992). In general, these models have argued that habitat fragmentation caused by glacial advances resulted in the isolation and subsequent divergence of populations in separate refugia. Given enough time for differentiation, populations would not hybridize following range expansions and contact during interglacial periods. Intergradation among different forms is often presumed to reflect secondary contact following glaciation (e.g., Barrowclough 1980b).

Within the past 25 years, a substantial body of information has accumulated concerning environments in the western United States during the late Pleistocene and early Holocene. In the more arid regions of the Great Basin and Southwest, plant macrofossil data obtained from packrat (*Neotoma*) middens have been especially informative (Betancourt et al. 1990). These records, collected at numerous scattered localities, have enabled workers to paint a clear picture of the past distribution of woodland habitats in the West, and to elucidate major ecological shifts that occurred during the Pleistocene-Holocene transition (see Porter 1983; H.E. Wright, Jr. 1983; R.F. Miller and Wigand 1994). At the peak of the last glaciation (approximately 18,000 yr BP), woodland vegetation dominated by junipers and/or pinyon covered most of the low to middle elevation areas from southern Nevada to southeastern California and northern Baja California (Wells and Jorgensen 1964; Wells and Berger 1967; T.J. King, Jr. 1976; Van Devender and Spaulding 1979; Wells 1979; McCarten and Van Devender 1988). Joshua trees and live oaks also occurred at certain zones in the woodland (Leskinen 1975; Wells 1979), with oaks becoming more prevalent toward the southeast in association with a gradient of increasing summer precipitation (Wells 1979). Although xerophytic plants intermingled with woodland species in the Southwest (Wells 1966; J.E. King and Van Devender 1977), desert vegetation was highly localized (Axelrod 1983) and limited to the lowest elevations and driest sites (e.g., the Colorado River Valley). In contrast to present-day patterns, pinyon or pinyon-juniper woodland was clearly absent from the northern and central Great Basin during the last glaciation (Thompson and Mead 1982; Wells 1983). Rather, subalpine coniferous forest dominated the lowland areas, with shrub communities occurring locally in valley bottoms on alluvial substrates (Thompson and Mead 1982; Wells 1983). Climatic changes in the early Holocene (approximately 10,000-8,000 yr BP; Axelrod 1981; O.K. Davis et al. 1985) brought about major elevational and latitudinal shifts in vegetation (e.g., Van Devender 1977; Cole 1982), resulting in zonation patterns similar to those observed today. While pinyon-juniper woodland migrated to higher latitudes and elevations, the expansion of xerophytic shrubs led to widespread development of deserts. Although regional deserts first appeared during interglacial cycles (Axelrod 1979), increasing aridity in the Holocene led to a previously unparalleled expansion of deserts that now characterize much of western North America (Van Devender 1977; Van Devender and Spaulding 1979; Axelrod 1979, 1983).

In view of the cooler conditions and associated vegetation changes that occurred in the Great Basin during glacial periods, titmice must have occupied warmer areas of woodland to the south at those times. On the other hand, populations in southeastern California and the Southwest, which are currently restricted to disjunct desert mountain ranges, probably enjoyed a more continuous distribution because of the downward displacement of junipers and pinyon into lowland areas. These populations most likely extended westward to the southern Sierra Nevada and San Joaquin Valley, where relict stands of pinyon persist (Axelrod 1981). As in the Great Basin, titmice along the Pacific slope also must have been restricted southward because of cooler and wetter climates in northern and central California during the Late Wisconsinan (Axelrod 1981). Populations probably occurred patchily along the length of Baja California where, at higher elevations, the Sierra Madrean element of the Madro-Tertiary flora (Axelrod 1958) would have served as a corridor of dispersal for numerous avian taxa that are currently isolated in the mountains of the southern Cape District (J. Davis 1959).

Dispersal and Gene Flow

The segregation of sclerophyllous plant communities into modern vegetation analogs was heightened during the early to middle Holocene, when a shift toward warmer, drier climates resulted in the present distribution of oak or oak-pine woodlands along the Pacific slope and juniper or pinyon-juniper woodlands in the interior (Axelrod 1981; O.K. Davis et al. 1985). This trend peaked during the Xerothermic (= Hypsithermal; see Pielou 1991:Fig. 1.6), with cooler and wetter climates characterizing the last 3,000 years (Axelrod 1981; O.K. Davis et al. 1985). Although western titmouse populations undoubtedly expanded their ranges during interglacial cycles, Holocene changes in temperature, precipitation, and vegetation would have enabled their northward dispersal into currently occupied areas. These movements inevitably resulted in increased gene flow among populations, especially along the Pacific slope, where stands of woodland are more continuous compared to the fragmented habitats in the interior.

Estimates of gene flow for *Parus inornatus* based on Slatkin's (1981, 1985b) methods were fairly high, with values on the order of those reported for birds that migrate and disperse long distances (e.g., Zink 1986; Rockwell and Barrowclough 1987; N.K. Johnson and Marten 1988, 1992; Burson 1990). Studies of other sedentary birds (e.g., *Callipepla californica*, Zink et al. 1987) also have revealed similar levels of gene flow. Although estimates calculated from F_{ST} (S. Wright 1951) were generally lower, they nonetheless indicated moderate gene flow among titmouse populations. These results were surprising, given the strict sedentary behavior of titmice once territories are established (Price 1936; Dixon 1949). Despite personal field impressions that interior birds are more vagile, there were no detectable differences in gene flow among Pacific slope and interior populations. However, the greater allozymic fragmentation (higher F_{ST}) in the interior implies lower gene flow there compared to the Pacific slope.

The importance of gene flow in constraining phenotypic or genetic differentiation has been the subject of considerable debate (e.g., Ehrlich and Raven 1969; Slatkin 1985b, 1987). One problem is the difficulty of obtaining direct estimates of gene flow, particularly for highly mobile organisms such as birds (Rockwell and Barrowclough 1987; but see Moore and Dolbeer 1989). Second, indirect estimates may reflect historical processes rather than current patterns of genetic exchange (Larson et al. 1984). Thus, fairly large values of *Nm* may be obtained for populations that are clearly isolated genetically from other conspecific populations (e.g., plethodontid salamanders, Larson et al. 1984; *Callipepla californica*, Zink et al. 1987). In addition, estimates of gene flow based on protein polymorphisms (S. Wright's F_{ST}) have been shown to overestimate levels of genetic exchange compared to mtDNA markers (Mitton 1994). Methods for estimating gene flow using mtDNA sequences are likely to reveal relatively small levels of *Nm* if the number of individuals sequenced from each population is ≥ 10 (Slatkin 1989). Whereas recent post-glacial dispersal and gene flow may account for the moderate to high estimates of *Nm* obtained for titmice in this study, current rates of gene exchange are probably substantially lower (especially in the interior). The strong genetic divergence between Pacific slope and interior populations reveals limited gene flow *between* regions.

Calibration of Molecular Divergences

Using the prevailing figure of 2% sequence divergence per million years (Brown et al. 1979; Shields and Wilson 1987b), Gill and Slikas (1992) proposed that *P. inornatus* split from *P. bicolor* during the "early Pleistocene" (i.e., 1.8 to 2 million years ago [MYA]). However, their actual estimates calculated from shared restriction fragments ($p[F] = 0.052$ to 0.057) and shared restriction sites ($p[S] = 0.071$ to 0.086) suggest earlier dates of 2.6 to 4.3 MYA. Similarly, the average percentage difference in cytochrome *b* sequences between these taxa (8.2%) indicates a divergence time of 4.1 MYA. An independent estimate of divergence time can be calculated from Nei's (1978) genetic distance (*D*). Marten and Johnson (1986) modified the calibration presented by Gutiérrez et al. (1983) for a fossil galliform (*Cyrtonix cooki*), in which they gave a conversion rate equivalent to $t = 19.7 \times 10^6 D$ (where *t* is time since divergence and *D* is Nei's [1978] genetic distance). From this formula, the two species of crested titmice appear to have diverged significantly later (only 1.3 MYA) than is indicated by the mtDNA data.

A similar discrepancy is obtained when estimating dates of divergence between Pacific slope and interior forms of *P. inornatus*. Whereas the mtDNA calibration suggests an early Pleistocene split dated at approximately 1.9 MYA, the allozyme data indicate a more recent divergence (average $D = 0.00327$ and $t = 64{,}400$ yr BP for all pairwise comparisons between these regions). Fairly recent divergence also is estimated for different subsets of coastal and interior populations, e.g.: Pacific slope vs. Warner and Northeastern California samples, average $D = 0.00996$, $t = 196{,}200$ yr BP; Pacific slope vs. Eastern Mojave samples, average $D = 0.0005$, $t = 9850$ yr BP. Interestingly, this timing of the split between coastal and Eastern Mojave titmice coincides precisely with the regional expansion of Southwestern deserts during the Pleistocene/Holocene transition.

Despite persistent attempts by systematists to estimate dates of divergence based on existing calibrations, such estimates are fraught with uncertainty (see discussion by Avise 1994:100-109). To begin with, difficulties in dating fossils undermine the reliability of calibrations based on either mtDNA or nuclear divergences (e.g., see Marten and Johnson 1986). Even if fossil dates are accurate, all calibrations assume the presence of a molecular clock (J.P. Thorpe 1982), a notion subject to considerable controversy (e.g., Gillespie 1986). The concept of a molecular clock further assumes that patterns of genetic divergence conform to the neutral, mutation-drift model (Barrowclough et al. 1985; Kimura 1979, 1982). Although the mtDNA calibration for primates (Brown et al. 1979) apparently holds true for geese as well (Shields and Wilson 1987a, 1987b), it may be inapplicable to songbirds or other groups with faster rates of evolution (Britten 1986; Bermingham et al. 1992). Within birds, evolutionary rates are suspected to differ between lineages of closely related taxa (e.g., Peterson 1992).

Because rates of change vary between and within different mitochondrial genes (e.g., Bibb et al. 1981; Edwards et al. 1991; Irwin et al. 1991; Cicero and Johnson 1995), calibrations of mtDNA divergence should be applied only to studies with comparable data sets. For example, calibration of the divergence between genera of geese (*Anser* vs. *Branta*, Shields and Wilson 1987b) was based on restriction enzyme analysis, which reflects an average rate of evolution over the entire mtDNA molecule. Therefore, this calibration is useful for sequence data only if estimates of sequence divergence are assumed to represent "average" differences. Likewise, estimated rates of sequence change for different taxonomic levels will be affected by the proportion of transitions versus transversions or silent versus replacement changes at different codon positions in the data set (Irwin et al. 1991:Fig. 3). The wide disparity in evolutionary rates among mitochondrial and nuclear genes implies the lack of a generalized molecular clock (Vawter and Brown 1986).

Model for Diversification in the *Parus inornatus* Complex and the Timing of Phylogenetic Events

Despite the difficulty of calibrating genetic distances, both the molecular and paleobotanical data yield insight into the probable timing of phylogenetic events in the *P. inornatus* complex. Allozymes and mtDNA indicate that the ancestor of *P. inornatus* diverged from a *P. bicolor* stock in the middle to late Pliocene or early Pleistocene. These dates correspond generally to the drastic changes that occurred in the Madro-Tertiary flora in response to increasing summer aridity and decreasing winter temperatures. Whereas ancestral populations probably inhabited the rich Madro-Tertiary flora of western North America during the Miocene, pre-*inornatus* evolved in conjunction with increased segregation and impoverishment of sclerophyllous woodland in the Pliocene and early Pleistocene. Populations of *P. inornatus* undoubtedly were confined to warmer regions in the south during periods of glaciation, although they probably shifted northward in interglacial times to other areas of favorable climate and habitat. While pioneer birds spread north along the Coast Ranges and both sides of the Sierra Nevada-Cascade mountains, they became increasingly divergent as a result of isolation by the montane barrier. Titmice along the Pacific slope adapted to oak or oak-pine woodlands found in areas of Mediterranean climate, whereas those in the interior evolved responses to juniper or pinyon-juniper woodlands occupying moister sites in cold winter climates. Furthermore, coastal populations in central and southern California experienced different selective regimes that coincided with the sharp floristic break caused by uplift of the Transverse Ranges (Axelrod 1958). Increases in temperature and aridity at the Pleistocene/Holocene transition, with associated changes in the distribution of woodland vegetation, enabled birds to continue their northward spread into central and northern portions of the current range. At the same time, populations in the Great Basin and Southwest became separated as lowland deserts expanded regionally and woodland was restricted to higher elevations. Although desert vegetation also expanded during interglacial cycles (Axelrod 1979), the peak regional development of deserts in the Southwest during the early to middle Holocene (Van Devender and Spaulding 1979; Axelrod 1983) undoubtedly severed any remaining link between interior and Pacific slope populations of titmice in southern California. Increasing aridity and expansion of desert environments also explains the isolation of titmice at the southern end of Baja California, where relict stands of Sierra Madrean woodland persist in the highlands (J. Davis 1959).

The presence of coastally derived titmice in northern California and southwestern Oregon probably has a more recent origin dating to the Xerothermic, when semi-arid woodland as well as chaparral (Detling 1961) spread at the expense of cold coniferous forest near Mount Shasta and in the Siskiyou Mountains. The phylogeographic pattern of mtDNA haplotypes indicates that dispersing individuals originated from source populations in the Coast Range rather than on the west slope of the Sierra Nevada. Such dispersal, combined with the appearance of western juniper on the Modoc Plateau at approximately the same time (R.F. Miller and Wigand 1994), would have created an opportunity for secondary contact between previously isolated, and already divergent, populations of coastal and interior titmice. Several lines of evidence (e.g., fire scars, tree age-class ratios, historical documents), however, indicate that prehistoric woodlands were "open, sparse, and savannah-like," and that the distribution and density of woodlands increased dramatically following European settlement (R.F. Miller and Wigand 1994:468). In fact, "the majority of present-day woodlands [in northeastern California and southeastern Oregon] are less than 100 years old" (ibid.:469). Thus, although Pacific slope and interior populations may have contacted intermittently since the Xerothermic, current patterns of distribution on the Modoc Plateau appear to be much more recent.

The above model of differentiation is consistent with observed patterns of genetic divergence. In addition to explaining the strong differentiation between titmice from the Pacific slope and interior (especially the northwestern Great Basin), it also accounts for (1) the low heterozygosity at Medford, which probably reflects a recent founding event dating to the Xerothermic; (2) the phenotypic and mtDNA break between coastal populations in central and southern California; and (3) the genetic similarity between coastal and interior titmice from southern and southeastern California, respectively. Allozymic and mtDNA data from the Eastern Mojave provide strong evidence of past genetic input from both the Pacific slope and interior during the evolutionary history of this population. In light of its current isolation, gene flow into this population from the two sources most likely occurred during glacial epochs when woodland habitat was more continuous. Support for post-Pleistocene isolation of titmice in this region is provided by the calibration of Nei's D between populations from the Pacific slope and Eastern Mojave ($t = 9850$ yr BP), which coincides with the timing of regional desertification in the Southwest during the early Holocene. Such vicariant events, coupled with interglacial and post-glacial dispersal, probably account for the current patterns of distribution and phylogenetic relationships in titmice in western North America.

TAXONOMIC PHILOSOPHY AND RECOMMENDATIONS

As stated in the introduction, the null hypothesis of this study was that populations of titmice are invariant in all characters examined. This approach enabled me to analyze patterns of variation without being constrained by previous views concerning relationships and taxonomy. Inter-population differences were evaluated in the framework of current subspecies boundaries. The following section presents recommendations for a taxonomic revision of the *P. inornatus* complex based on geographic differences in phenotypic, genetic, vocal, and ecologic traits, and discusses the rationale behind these decisions.

Sibling Species of Titmice

A persistent debate in systematic ornithology pits advocates of the biological species concept (Dobzhansky 1937; Mayr 1963) against proponents of the phylogenetic species concept (Cracraft 1983; McKitrick and Zink 1988). Recently, Avise and Ball (1990) attempted to incorporate strengths of both definitions using concordance principles, i.e., species should be recognized if *multiple* independent genetic characters show phylogenetic concordance that reflects *intrinsic* barriers to reproduction (Avise 1994; italics mine). Thus, reconstructed phylogenies should represent species trees rather than gene trees, where sufficient time has elapsed for the lineage sorting process to produce monophyletic sister taxa rather than paraphyletic or polyphyletic lineages (see Avise 1994:126-138).

The concept of concordance is crucial to interpreting patterns of variation in titmice. In particular, the fact that Pacific slope and interior populations are distinguishable based on morphometric, colorimetric, allozymic, mtDNA, *and* ecologic traits indicates that they represent integrated evolutionary units. Furthermore, preliminary data on vocalizations, which revealed fundamental differences in both the structure and tempo of song repertoires between coastal and interior titmice, also point to intrinsic reproductive isolation. The importance of vocalizations in differentiating sibling or cryptic species has been documented for a number of other avian taxa (e.g., *Empidonax* [N.K. Johnson 1980]; *Polioptila* [Atwood 1988]; *Loxia* [Groth 1988, 1993]).

Although such strong concordance among character types supports the treatment of Pacific slope and interior populations as sibling species, at least in the phylogenetic sense (Cracraft 1983), recent secondary contact and *limited* gene flow on the Modoc Plateau indicate that the two clades are not completely isolated reproductively. Nonetheless, I do not view this as a problem with regard to the biologic species concept. To begin with, this concept allows for limited genetic exhange (Dobzhansky 1937), and numerous biologic species, including examples within *Parus* (e.g., *P. atricapillus* and *P. carolinensis*, Robbins et al. 1986; Braun and Robbins 1986), hybridize in the wild. Furthermore, the relatively high F_{ST} value (0.147) among populations of titmice in that region, combined with the sharp break in mtDNA haplotypes, provides strong evidence for a barrier to gene flow. Such genetic differences, which are accompanied by morphologic and colorimetric changes, could not be maintained in the face of extensive hybridization and introgression. The tenuous nature of this narrow, putative contact zone is highlighted by the patchy distribution and relatively low density of titmice on the Modoc Plateau. Given such conditions, fluctuations in the distribution and density of juniper (R.F. Miller and Wigand 1994) in that region would create only intermittent opportunities for hybridization between differentiated populations of titmice.

Despite limited gene flow in northern California, Pacific slope and interior populations clearly form two monophyletic clades that have speciated under different climatic and ecologic regimes. Consequently, I regard them as full sibling species (sensu Amadon and Short 1992) and recommend formal adoption of the following names by the American Ornithologists' Union Committee on Classification and Nomenclature: *P. inornatus*, "Oak Titmouse," for Pacific slope populations; and *P. ridgwayi*, "Juniper Titmouse," for interior populations. These vernacular names appropriately describe the distinct habitat preferences of the two forms. Although some exceptions occur, Pacific slope titmice are strongly associated with oak-dominated woodlands, whereas interior titmice prefer more arid woodlands dominated by juniper.

Subspecies Boundaries

Controversial arguments surrounding the validity of subspecies also persist in systematics and taxonomy (E.O. Wilson and Brown 1953; Wiens 1982; Ball and Avise 1992; Patten et al. 1995). Nonetheless, I support the continued use of subspecies *if* the named forms reflect discrete and objectively defined populations or groups of populations (see Barrowclough 1982; N.K. Johnson 1982; O'Neill 1982; Parkes 1982; Storer 1982). Thus, subspecific names should not be ascribed to populations that merely represent ends of a gradual cline. Genetic units may be divisible into smaller entities if such populations are diagnosable based on morphologic and/or colorimetric traits (Patton and Smith 1990).

Decisions regarding subspecies boundaries of titmice were based on several considerations: univariate and multivariate patterns of differentiation in size and/or color; patterns of variation in allelic frequencies and/or mtDNA haplotypes; and geographic distribution as well as habitat preference. Because tissues were unavailable from Baja California, the taxonomy of those populations was based entirely on phenetic characteristics and distributional occurrence. Analyses of size variation revealed two major groupings of populations within the range of the "Oak Titmouse": (I) small birds with short to medium-length toes, which fall into the currently recognized subspecies *sequestratus*, *inornatus*, *kernensis*, and *cineraceus*; and (II) moderately large birds with relatively long toes and larger bills (*transpositus*, *mohavensis*, *affabilis*). These groups also differed significantly in color, the latter (except for *mohavensis*) being characterized by darker dorsal and ventral plumage. Although specimens of *kernensis* generally were larger than those to the north along the western slope of the Sierra Nevada, they did not differ substantially in size or color from most other populations of titmice comprising group I. Patterns of allozyme variation likewise failed to reveal any unique characteristics for *kernensis*. Because the evidence does not support recognition of this subspecies, I recommend merging populations of *kernensis* with those comprising the nominate form *P. i. inornatus*. Populations of *P. i. inornatus* along the foothills of the Sierra Nevada, which differed from other *inornatus* both in size and mtDNA haplotype, were not sufficiently distinct to warrant elevation to subspecies status. In contrast, titmice of the highly isolated form *cineraceus* in southern Baja California, which presumably are most closely related to birds from northern Baja California (i.e., *affabilis*), deserve recognition because of their much smaller size, and generally paler plumage, compared to the closest populations farther north.

A more complicated situation arises with regard to the status of *sequestratus* in northern California and southwestern Oregon. These populations, which are isolated from other Pacific slope titmice to the south by coniferous forest near Mt. Shasta, resemble *inornatus* in both size and color. However, secondary contact with interior titmice has led to allozymic introgression, particularly at the LGG locus. Use of cytochrome *b* sequences as molecular markers (Mack et al. 1986) shows that titmice of the coastal form *sequestratus* range at least as far east as Mt. Dome (sample area 3) on the Modoc Plateau. The purest examples of *sequestratus* occur in the Rogue River Valley (i.e., Medford [1]) of Oregon, where the type specimens were collected and described. Titmice there are smaller, on average, than other "*sequestratus*" from California, and do not show any sign of gene flow from the interior. Furthermore, these populations are more strictly associated with oaks than related populations in northern California, which inhabit pure oak woodland, oaks mixed with junipers, or pure junipers. Although secondary contact has led to the infusion of interior allozymic and ecologic traits on the Modoc Plateau, *sequestratus* from the type locality and surrounding region are not sufficiently distinct from *inornatus* to warrant separate status. Thus, I favor the conservative stance of merging all coastal populations in southwestern Oregon and northern California, at least as far east as Mt. Dome, into *inornatus*.

Titmice in southwestern California (*transpositus*, group II) varied clinally from Santa Barbara and Ventura counties, where they intergrade with *inornatus* (group I), southward to the west slope of the Sierra San Pedro Mártir in northern Baja California. Despite evidence of intergradation between *transpositus* and *inornatus*, populations of these two forms differed sharply in several morphometric and colorimetric traits. In contrast, there was no clear break in most characters at the taxonomic boundary between *transpositus* in southern San Diego County and *affabilis* in northern Baja California. While multidimensional scaling of size showed slight separation between females of *affabilis* and *transpositus*, the plot of males revealed substantial overlap between these forms. Likewise, the quantitative color data also failed to separate *transpositus* from *affabilis*. Accordingly, I do not recognize these populations as distinct subspecies. I recommend combining them under the name *affabilis*, which was established prior to *transpositus*.

A.H. Miller (1946) originally described *mohavensis*, which is endemic to the Little San

Bernardino Mountains, primarily on the basis of qualitative differences in plumage color. Given the data on allozymes and mtDNA variation, *mohavensis* is clearly derived from Pacific slope populations to the west. Nonetheless, the quantitative appraisal of color variation revealed significant differences between these populations and supported Miller's basis for subspecific recognition of *mohavensis*. Whereas specimens from the Little San Bernardino Mountains showed some resemblance in color to other forms of "Oak Titmice," they clearly varied from the nearest sampled populations (= "*affabilis*" from the San Bernardino and San Jacinto mountains) in brightness and purity of both the dorsum and breast. The significantly grayer plumage of *mohavensis* compared to "*affabilis*" reflects adaptation to a more arid environment, in which the preferred habitat consists of juniper or pinyon-juniper woodland (mixed with Joshua trees) rather than oak woodland. Size differences between "*affabilis*" and *mohavensis* were more subtle, although both males and females showed some separation in the multidimensional scaling analyses of size. Miller noted that populations of *mohavensis* and "*affabilis*" may intergrade on the east slope of the San Bernardino Mountains, where they occupy a similar habitat of junipers mixed with Joshua trees. Nonetheless, I favor the continued recognition of the subspecies *mohavensis* for titmice in the Little San Bernardino Mountains.

In addition to groups I and II above, interior populations of "Juniper Titmice" formed a third major grouping based on differences in size and color. These populations are currently ascribed to three subspecies: *zaleptus*, *ridgwayi*, and *plumbescens*. Although size generally declined from the more northerly interior populations toward the Southwest, univariate and multivariate analyses of size variation did not reveal any breaks corresponding to subspecies boundaries. Likewise, the forms overlapped significantly with regard to dorsal and ventral coloration. In contrast, the allozyme data clearly separated interior populations of titmice into two distinct groups (see Figs. 37-39): those from the western Great Basin (= currently recognized *zaleptus*) *plus* Southeastern Nevada (39) and the Ruby Mountains (40); and those comprising other western and southwestern populations (= most of *ridgwayi* as well as *plumbescens*). Despite the lack of "visible" differentiation between these two groups (see Amadon and Short 1992:14), I recommend that they be recognized as *P. ridgwayi zaleptus* and *P. r. ridgwayi*, respectively. It is important to note that the boundary of *P. r. zaleptus* does not correspond with the previously established limit of *zalepetus* in the western Great Basin.

The sample from the Eastern Mojave (37) requires special discussion because of its unique characteristics. Although titmice from this area are clearly interior phenotypically, geographically, and ecologically, patterns of allozyme variation showed strong similarity to the "Oak Titmouse" clade. On the other hand, an unusual allele at the ADA locus provided an important clue linking samples from the Eastern Mojave and the western Great Basin. The presence of both coastal and interior mtDNA sequence haplotypes in the Eastern Mojave further complicates the picture. These genetic results were completely unexpected and indicate mixed phylogenetic ancestry. The conflicting patterns exhibited by the phenotypic and genetic data present a problem for the taxonomy of titmice in the

Eastern Mojave. Nonetheless, logic dictates that this population be assigned to an interior subspecies based on phenotypic, biogeographic, and habitat considerations. Therefore, I include it with other western populations of "Juniper Titmice" (*"P. r. zaleptus"*).

Summary of Taxonomic Revision

According to my study, populations of crested titmice in the western United States and Baja California, Mexico, form a superspecies comprised of two "allospecies" or sibling species (sensu Amadon and Short 1992), *Parus inornatus* and *P. ridgwayi*. On a macrogeographic scale, these species are monophyletic, cohesive evolutionary units with distinctive phenotypic, genetic, vocal, and ecologic traits. Each species is further subdivided geographically into two or more subspecies that have differentiated sufficiently from other forms to warrant recognition as separate entities. The following hierarchy summarizes the results of taxonomic decisions presented above. Revised taxonomic units are illustrated in Fig. 45. Further discussions of these taxa, along with nomenclatural synonymies, are presented in Appendix B.

Parus inornatus ("Oak Titmouse"):	*Parus inornatus inornatus*
	Parus inornatus mohavensis
	Parus inornatus affabilis
	Parus inornatus cineraceus
Parus ridgwayi ("Juniper Titmouse"):	*Parus ridgwayi zaleptus*
	Parus ridgwayi ridgwayi

SUMMARY

This study presents the results of a comprehensive investigation of macrogeographic variation in the Plain Titmouse (*Parus inornatus*) complex throughout its range in western North America. Currently regarded as a single species, the primary objective was to assess *population-level* patterns of differentiation in *P. inornatus* using modern systematic techniques applied to morphometric, colorimetric, allozymic, mtDNA, and vocal characters. While subspecific boundaries were ignored for the purposes of data collection and analysis, patterns of variation were interpreted in the framework of existing subspecies to test their validity. Such patterns also were used to infer historical processes of divergence in the species complex and to assess the phylogenetic relationships of populations.

The range of *P. inornatus* is subdivided into several disjunct groups of populations that are separated from each other by barriers of unsuitable habitat. These groups vary strikingly in their habitat preference, population density, and extent of fragmentation among populations. Morphometric and colorimetric traits exhibit clinal, disjunct, or

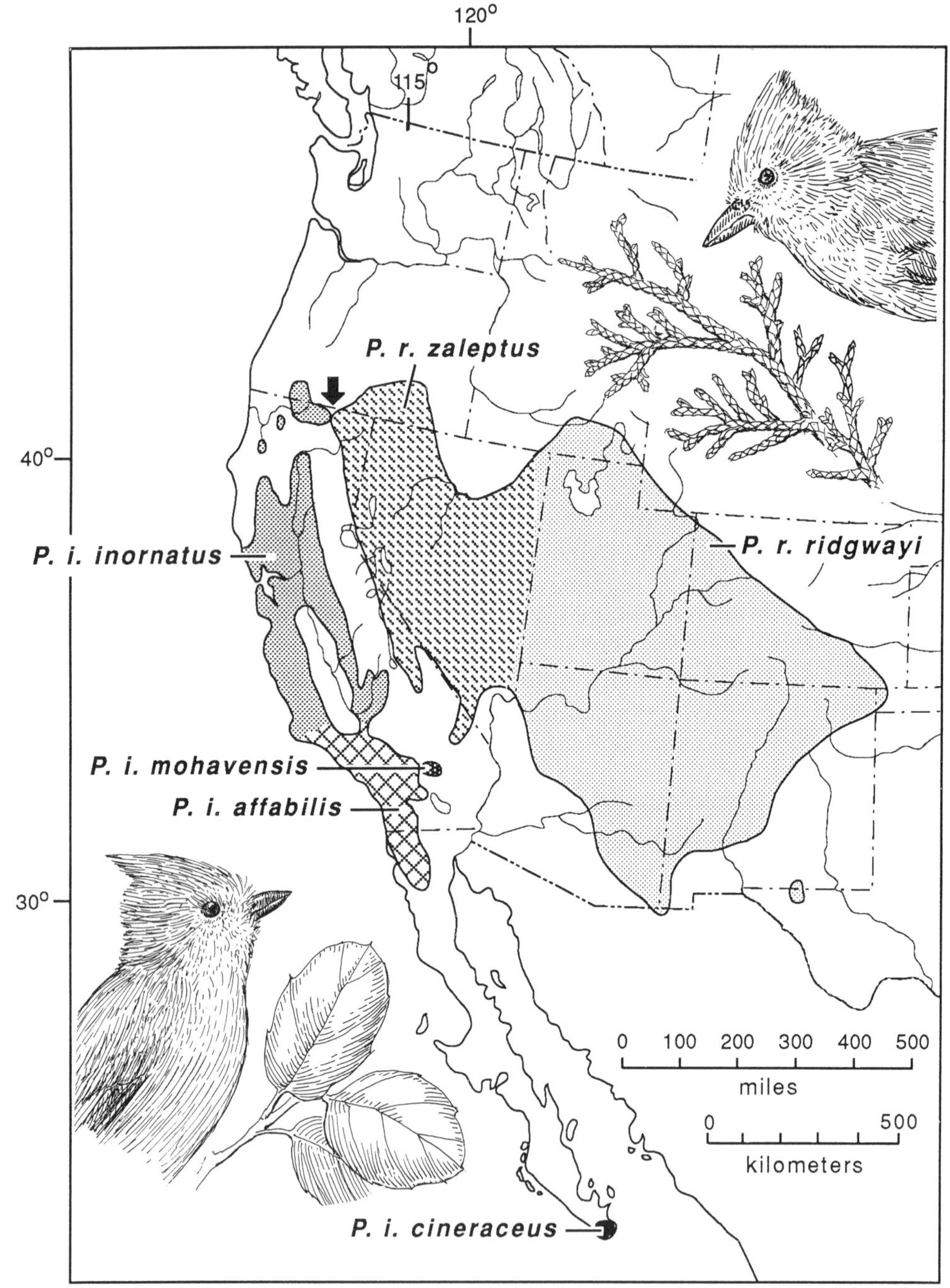

120°

115°

P. r. zaleptus

40°

P. i. inornatus

P. r. ridgwayi

P. i. mohavensis

P. i. affabilis

30°

0   100 200 300 400 500
miles

0   500
kilometers

P. i. cineraceus

FIGURE 45. Generalized distribution of titmice in western North America based on taxonomic changes recommended in this study. Two sibling species are recognized: *Parus inornatus* (Oak Titmouse), which extends along the Pacific slope from southwestern Oregon to southern Baja California; and *Parus ridgwayi* (Juniper Titmouse), which occurs from the western Great Basin to the Rocky Mountains and Southwest. The approximate ranges of four suspecies of *P. inornatus* and two subspecies of *P. ridgwayi* are indicated. An isolated population (not shown) of as-yet-undetermined affinity occurs near Fall River Mills, Shasta Co., California, about half-way between the known limits of *P. i. inornatus* and *P. r. zaleptus*. The arrow points to a narrow, tenuous secondary contact zone on the Modoc Plateau in northern California.

mosaic patterns of variation. Pacific slope populations, which are associated primarily with oak-dominated woodlands or savannas, are comprised of: (I) small birds with small bills and short toes (presently recognized subspecies *sequestratus*, *inornatus*, *kernensis*, and *cineraceus*); and (II) moderately large birds with long toes and larger, heavier bills (*transpositus*, *affabilis*, and *mohavensis*). A third grouping, characterized by moderately large to large birds with large bills and short toes (*zaleptus*, *ridgwayi*, *plumbescens*), inhabits more arid juniper or pinyon-juniper woodlands ranging from the western Great Basin to the Rocky Mountains and Southwest. Interior titmice are significantly paler and grayer (i.e., higher brightness and lower purity of color) than those from the Pacific slope. Birds from southern California and northern Baja California (size group II) differ from other coastal titmice (group I) in having lower values for brightness of the back and breast, indicating an overall darker plumage. Titmice from group I are characterized by a whiter breast that strongly separates them in multidimensional space. Mantel tests showed significant, broad-scale geographic congruence between patterns of size and color variation. Both morphometric and colorimetric traits are highly correlated with several environmental variables, including elevation, longitude, average minimum and mean temperatures during December and January, seasonal temperature differences, mean August precipitation, and mean precipitation during December and January.

Levels of genetic variability within populations were low compared to other avian taxa, averaging 0.025 for observed heterozygosity and 13.81% for the percentage of loci that were polymorphic. Variability was greater in populations from the interior than from the Pacific slope. S. Wright's (1978) F_{ST} for all samples combined was 0.106, which is relatively high for conspecific populations of birds and indicates a definite departure from panmixia. As expected because of their more fragmented distribution, interior populations exhibited higher allozymic subdivision ($F_{ST} = 0.072$) compared to Pacific slope populations ($F_{ST} = 0.059$). The highest F_{ST} values and genetic distances occurred between coastal and interior populations from southern Oregon and northern California ($F_{ST} = 0.147$; Nei's $D = 0.013$). Clustering and ordination techniques applied to the allozyme data clearly differentiate Pacific slope versus interior titmice and further subdivide interior populations into two geographic groups. Electrophoretically, birds from southeastern California are closer to those from the Pacific slope than to other other interior populations. A population of mixed allozymic composition occurs on the Modoc Plateau in northern California. Different estimates of gene flow indicate moderate to high levels of exchange among populations, which is surprising given the sedentary nature of the species. Mantel tests indicate that the patterns of allozymic variation are concordant with those exhibited by size but not by color.

Cytochrome *b* sequences revealed five distinct mtDNA haplotypes in *P. inornatus*. Two haplotypes were confined to Pacific slope populations, two occurred only in the interior, and one was shared by Pacific slope and interior populations from southern and southeastern California. A population from the eastern Mojave Desert was polymorphic for both coastal and interior haplotypes. The average percentage nucleotide difference varied from 0.7% among Pacific slope haplotypes to 4.0% between the two regions. Percentage sequence divergence was higher in outgroup comparisons, ranging from 8.2% (*P. inornatus* vs. *P. bicolor*) to 11.1% (*P. bicolor* vs. *P. gambeli*) and 11.3% (*P. inornatus* vs. *P. gambeli*). Mantel tests revealed a strong association between mtDNA and allozyme distances among populations of *P. inornatus*. According to the mtDNA data, interior populations of *P. inornatus* are more similar to *P. bicolor* than Pacific slope populations.

Preliminary analysis of vocal data revealed that Pacific slope and interior titmice differ in the fundamental structure and tempo of their songs, a finding that has critical implications for the reproductive isolation of those populations. Although no song types were shared among titmice from different Pacific slope samples, males from geographically distant interior populations shared three song types. Several song types were shared by males from *within* each sample.

Results of the genetic analyses, coupled with paleobotanical data, are used to develop hypotheses for the differentiation of titmice in western North America. The ancestor of *P. inornatus* probably diverged from a *P. bicolor* stock in the middle to late Pliocene and early Pleistocene, when increasing summer aridity and decreasing winter temperatures led to impoverishment and segregation of sclerophyllous woodland. Northward dispersal of pioneer populations during interglacial and post-glacial times resulted in isolation and subsequent divergence of titmice on both sides of the Sierra Nevada-Cascade mountains. Regional desertification in southeastern California during the Pleistocene/Holocene transition severed a late glacial connection between Pacific slope and interior populations. Evidence for gene flow between coastal and interior titmice on the Modoc Plateau in northern California, coupled with information on population density and distribution as well as juniper macrofossils, indicate recent but tenuous secondary contact.

Broad-scale geographic congruence exhibited by morphometric, colorimetric, allozymic, mtDNA, vocal, and ecologic traits indicate that Pacific slope and interior titmice form monophyletic clades with different evolutionary trajectories. Both the phylogenetic and biologic species concepts support treatment of these groups as sibling species within a superspecies complex. The names *Parus inornatus* ("Oak Titmouse") and *P. ridgwayi* ("Juniper Titmouse") are recommended for coastal and interior forms, respectively. Other taxonomic decisions at the subspecies level include: merging *kernensis* and *sequestratus* into nominate *inornatus*; lumping *transpositus* and *affabilis* into one subspecies (*affabilis*); merging *plumbescens* into *P. r. ridgwayi*; and extending the range of *P. r. zaleptus*, based on allelic frequencies, eastward to approximately the Nevada-Utah border. As in most studies of cryptic species, the evolutionary status of Pacific slope and interior titmice as sibling species would have remained unexposed without a thorough geographic survey using modern systematic techniques.

Appendix A

Sample Areas and Localities of Specimens

Specific localities of titmouse specimens examined in this study are given below for each sample area. Sample areas are arranged numerically according to Figs. 3-4 and Table 2. Localities are written exactly as indicated on specimen tags; subtle variations or changes in place names (indicated by "="), and/or additional locality or elevational information provided on some specimen tags, are given in brackets ([ ]). Elevations were combined for specific localities to indicate the elevational range at that site. The total number of specimens examined per locality is indicated by the letter "T." The actual number of specimens *analyzed* from each locality is indicated as follows: "M," males measured for morphometric analyses; "F," females measured for morphometric analyses; "C," males and females examined for colorimetric analyses; and "E," tissue samples analyzed for allozyme variation. Samples analyzed for mtDNA sequence variation are a subset of those denoted by "E." See Table 2 for complete sample sizes used in each set of analyses. Specimens examined electrophoretically ("E") were all collected during the course of this study; other specimens were obtained from existing museum collections. Localities of specimens collected during this study and representing new distributional records are highlighted in boldface.

MEDFORD (1)

OREGON: Jackson County: Ashland (T=5, M=3, F=2, C=4); Brownsboro (T=5, M=1); 6 mi E Brownsboro, 2000 ft (T=1, C=1); 8 mi N Eagle Point (T=1, F=1); 7 mi N Eagle Point (T=2, M=2); 6 mi N Eagle Point (T=6, M=3, F=3, C=4); 5 mi N Eagle Point (T=7, M=4, F=1); near Lick Cr., 1700 ft, 1 mi N & 6.5 mi E Eagle Point (T=14, M=6, F=6, E=14); NE of Upper Table Rock, 1500 ft, 1 mi N & 4 mi W Eagle Point (T=1, M=1, E=1); Eagle Point (T=13, M=8, F=2, C=6); Medford (T=6, M=3, F=2); Payne Cliffs (T=5); Sam's Valley (T=1, M=1). **Josephine County**: Grant's Pass (T=3, F=3, C=2); 3 mi SW Merlin (T=3, M=2, F=1).

SISKIYOU (2)

CALIFORNIA: **Siskiyou County**: Beswick (T=3, M=3, C=3); 2500 ft, near Bogus (T=4, M=1, F=3, C=4); Little Bogus Cr., 4 mi NE Ager, 2600 ft (T=5, M=1, F=2, C=5); NW end Copco Lake, 3000 ft, 1.25 mi N & 1 mi E Daggett Mtn. (T=11, M=3, F=3, E=6); Hornbrook (T=1, M=1); Juniper Flat, 3200 ft, 2.25 mi N & 4.25 mi E Lake Shastina dam (T=11, M=5, F=2, E=11); S slope Little Shasta R. canyon, 3800 ft, 2 mi N & 12 mi E Montague (T=3, F=1, E=3); 3 mi W Secret Spring Mtn., 3000 ft (T=13, M=6, F=5, E=11); caves, 11 mi NE Weed, 3600 ft (T=1, M=1).

MT. DOME (3)

CALIFORNIA: **Siskiyou County: W of Box Canyon, 4200 ft, 1.75 mi S & 2 mi W Mt. Dome** (T=1, M=1, E=1); **NW of Box Canyon, 4300 ft, 0.5 mi S & 1.5 mi W Mt. Dome** (T=11, M=6, F=5, E=11); **near Dorris Brownwell Rd., 4400 ft, 1 mi S & 1.75 mi E Dorris** (T=2, M=1, F=1, E=2); **Modoc Gulch, 4400 ft, 2.25 mi S & 2.25 mi E Dorris** (T=3, M=1, E=3); **S of Red Rock Valley, 4400 ft, 1 mi N & 4.25 mi E Cedar Mtn.** (T=8, M=6, F=2, E=8).

NORTH SACRAMENTO VALLEY (4)

CALIFORNIA: **Shasta County**: Anderson (T=1, M=1); Backbone Ridge, 1200 ft, 3 mi N & 7 mi E Project City (T=17, M=8, F=6, E=16); Baird Station (T=1, M=1); Cottonwood, near Redding (T=1, M=1); Redding (T=1, M=1); Tower House, 1264-1268 ft (T=2, M=1, F=1). **Tehama County:**   Battle Cr. (T=1, M=1); Battle Cr. bounding Tehama & Shasta cos. (T=1, M=1); mouth of Battle Cr., 350 ft (T=1, M=1); Paine [= Payne] Cr., 600 ft (T=7, M=3, F=3); Payne's Cr. (T=1, M=1); Dale's, 600 ft, on Payne's Cr. (T=3); Payne P.O. (T=1, F=1, C=1); Payne Cr. P.O., 1600 ft (T=3, F=1); 2800 ft, 8 mi E Paynes Cr. P.O. (T=1, F=1); 3500 ft, 11 mi E Payne Cr. P.O. (T=1, F=1); 6 mi NE Red Bluff, Antelope Canyon (T=1, F=1); 2 mi N Red Bluff, 350 ft (T=1); 300 ft, 1 mi NE Red Bluff (T=3, M=2, F=1); Red Bluff (T=19, M=9, F=10, C=2); 3 mi E Red Bluff (T=4, M=3, F=1); 4 mi E Red Bluff (T=2, F=2); Mill Cr., 2 mi NE Tehama, 260 ft (T=4, M=2).

CLEAR LAKE (5)

CALIFORNIA: **Colusa County**: 5 mi W Sites (T=4, M=3, F=1); Sites (T=11, M=3, F=4); Snow Mtn. (T=2, M=1, F=1); Stonyford (T=2, M=1, F=1). **Glenn County:** Winslow, 5 mi W Fruto, 700 ft (T=6, M=1). **Lake County:** Cache Cr. (T=2, M=1, F=1); Castle Hot Springs [=Castle Rock] (T=1, F=1); Clear Lake (T=1, M=1); 6 mi NW Clearlake Park, 1350 ft (T=1, M=1); Harbin Springs [near Middletown] (T=5, M=4, F=1); Lakeport (T=4, M=3, F=1); Lower Lake (T=1, M=1); 3 mi N & 2.25 mi E Pinnacle Rock (T=1, M=1, C=1); 3 mi N Upper Lake (T=1, F=1). **Mendocino County:** 6 mi E Covelo (T=1); 20 mi E Covelo (T=3, M=1); Hearst P.O. (T=3, M=1, F=2); Ukiah (T=7, M=4, C=1). **Sonoma County:** Russian R., 1.4 mi N Cloverdale (T=1, M=1); Preston (T=1). **Yolo County:** Davis Cr. drainage (near Davis Cr. Reservoir), 1600 ft, 1.25 mi N & 0.75 mi W Knoxville (T=17, M=13, F=4, E=17); Rumsey, 500 ft (T=1).

OROVILLE (6)

CALIFORNIA: **Butte County**: Butte Cr., 10 mi NW Biggs (T=2, F=2); Butte Cr., 4 mi SE Chico, 200 ft. (T=11, F=1); 10 mi NE Chico (T=1); 7 mi W Chico (T=1, F=1); Chico (T=3,

M=1); Enterprise (T=12, M=3, F=9, C=10); Hurlton (T=1, F=1); Chambers Ravine, 4 mi N Oroville, 600 ft. (T=6, M=2); Yankee Hill (T=1, F=1). **Nevada County:** Cherokee, W slope Sierra Nevada mts. (T=1, M=1); Grass Valley (T=3, M=2, F=1); Nevada [= Nevada City] (T=1). **Sutter County:** Marysville Buttes, 500 ft, 4 mi NW Sutter (T=1, F=1); Marysville Buttes, 500-800 ft, 3 mi NW Sutter (T=2, M=2); Marysville Buttes (T=1, M=1); West Butte (T=1). **Yuba County:** 7 mi N & 14 mi E Marysville, 400 ft (T=4, M=2, F=1, E=4); Marysville (T=2, M=1, F=1, C=2); Rackerby (T=2, M=2); Oat Hills, 1100 ft, 3.5 mi N & 0.5 mi W Smartville (T=17, M=6, F=3, E=10); Timbuctoo (T=1, M=1, C=1).

SAN FRANCISCO BAY AREA-NORTH (7)

CALIFORNIA: Marin County: Black's Mtn. (T=1); Maillard (T=2, M=1, F=1); Miller (T=1, M=1); Nicasio (T=5, M=3, F=2, C=2); San Geronimo (T=6, M=3, F=2); no locality (T=5, M=2, F=3). **Napa County:** 3 mi W Monticello dam (T=7, M=6, F=1); 4 mi SE Monticello (T=1, M=1); St. Helena (T=2, M=1, F=1, C=2); Yountville (T=2); no locality (T=3, M=1). **Solano County:** 8 mi W Vacaville, 700 ft (T=5). **Sonoma County:** Cotati (T=7, M=4, F=2); Eldridge (T=1, M=1); Freestone (T=1); Fulton (T=3, M=2, F=1); O'Connor Ranch, 3 mi NNE Kellogg (T=1, M=1); Petaluma (T=1, M=1); 1.5 mi W Petaluma (T=1, M=1); Santa Rosa (T=7, M=1, F=3, C=2); Sebastopol (T=7, M=2, F=5, C=3); Sonoma Mtn., 8 mi E Petaluma (T=2, M=1, F=1).

SAN FRANCISCO BAY AREA-SOUTH (8)

CALIFORNIA: Alameda County: Alameda (T=2, F=1); Berkeley (T=11, M=4, F=6, C=1); Hamilton Gulch, Berkeley (T=2, M=1, F=1); Dublin (T=1); Hayward [= Haywards] (T=18, M=8, F=8, C=3); Diamond, Oakland (T=1); 0.5 mi NW municipal golf links, 200 ft, Oakland (T=1, M=1); Oakland (T=11, M=3, F=4, C=1); San Leandro (T=1, F=1); Strawberry Canyon [600 ft] [Berkeley] (T=3, M=1); Sunol (T=5, M=1, F=1); La Costa Cr., 5 mi SE Sunol (T=1); Hoag's Canyon, Temescal [= Temescal] (T=3, F=3, C=2). **Contra Costa County:** 1 mi NW Alamo [275 ft] (T=2, M=2); 300 ft, 3 mi N Berkeley (T=1, M=1); Black Hills, 1000 ft, 3 mi N Orinda Village (T=1, M=1); Black Hills, 1020 ft, 2.6 mi N Orinda Village (T=1, F=1); Bollinger Cr., 3.25 mi SW Danville (T=2); Clayton, Curry Cr. Canyon (T=1); Diablo (T=1, M=1); Mt. Diablo (T=3, M=1, F=2); W side Mt. Diablo, 1750-2750 ft (T=2, M=1); lower slopes Mt. Diablo (T=1); 2.5 mi E Mt. Diablo (T=1, F=1); 3 mi SE Mt. Diablo (T=1, M=1); 1 mi N Lafayette [900 ft] (T=2, F=2); 1.5 mi W Lafayette (T=1, F=1); Lafayette (T=1, M=1); Martinez (T=1, M=1); 0.25 mi NE Saint Mary's College, Moraga Valley (T=1, F=1); Moraga Valley [5 mi NE Berkeley] (T=6, M=3, F=2); Nortonville [600 ft] (T=3, M=1, F=2, C=1); 1 mi NW Orinda store (T=1, M=1); 1.8 mi S & 1 mi E Orinda, 800 ft (T=1, C=1); Pinole Cr., 3 mi SE Pinole, 250 ft (T=2); Saranap, 1 mi W Walnut Creek (T=1, M=1); Aloha Farm, Walnut Creek (T=1, M=1); Walnut Creek (T=5, M=1, F=3, C=4); 2 mi SW Walnut Creek, 200 ft (T=5, M=1). **San Francisco County:** San Francisco (T=1, F=1). **San Mateo County:** Belmont (T=1, F=1, C=1); Jasper Ridge (T=1, F=1); La Honda (T=1, M=1); Menlo Park (T=6, M=2, F=4, C=2); Portola (T=7, M=4, F=3, C=5); Redwood City [= Redwood] (T=4, M=2, F=1, C=1); 2 mi S Redwood City (T=1); 0.75 mi S Golden Gate Nat'l. Cemetery, San Bruno (T=1, M=1); Sharon Estate, 1300 ft (T=1, M=1); "in foothills" (T=1, F=1); no locality (T=3, M=1, F=2). **Santa Clara County:** Alviso (T=2, M=1); Berryessa (T=6, M=2, F=2, C=3); Calaveras Valley (T=1, M=1); 4 mi S Evergreen (T=1, M=1); Los Gatos (T=7, M=3, F=3, C=1); Mayfield (T=1, M=1); near Mayfield (T=1, F=1); Palo Alto [40 ft] (T=50, M=30, F=16, C=10); near [vic.] Palo Alto (T=8, M=5, F=2); Palo Alto Gun

& Rod Club, 150 ft [= Palo Alto Shooting Club] (T=3, M=3); San Jose (T=4, M=2, F=1, C=2); Santa Clara (T=1); Stanford Univ. [Palo Alto] (T=7, M=3, F=2); Felt Lake, Stanford Univ. (T=1, F=1); hills near Stanford Univ. (T=1, M=1); Stevens Cr. (T=1, M=1); "near the bay" (T=1, M=1).

STOCKTON (9)
CALIFORNIA: San Joaquin County: Linden (T=1, M=1); 7 mi NE Lodi (T=1, C=1); Stockton (T=4, M=2, F=2); Sulfur Spring Gulch, 2500 ft, 9 mi S & 3 mi W Tracy (T=1, M=1); Tracy (T=2, M=2). **Stanislaus County:** Modesto (T=1, M=1); Oakdale (T=3).

MONTEREY (10)
CALIFORNIA: Monterey County: Arroyo Seco, 10 mi S Paraiso Springs (T=1, M=1); headwaters Big Cr., Lucia P.O. (T=4, M=1); Big Pines, 3700 ft, 9 mi W Jamesburg (T=3, M=1, F=2); N edge Big Pines, 3900 ft (T=1, F=1); Carmel (T=2); Carmel Valley (T=2); 12 mi up Carmel Valley [Los Ranchitos] (T=8, M=4, F=3, C=5); 2.7 mi E Chalk Pk. [Nacimiento R.] (T=2, M=1, F=1, C=2); 1.25 mi S Chalk Pk., 3000 ft (T=1); China Camp, 4500 ft, Santa Lucia Mts. (T=1, M=1); Church Cr., 2500 ft, Santa Lucia Mts. (T=1, M=1); 7 mi SE Del Monte (T=2, M=1); Haystack Hill, 2000 ft, Hastings Reservation, 6.5 mi S & 9.75 mi E Carmel Valley (T=1); Poison Oak Ridge, 2200 ft, Hastings Nat. Hist. Reservation, 0.25 mi N & 2 mi E Jamesburg (T=5, M=3, F=2, E=5); Hastings Nat. Hist. Reservation, 2600 ft, 0.25 mi S & 1.5 mi E Jamesburg (T=11, M=6, F=5, E=11); Hastings Reservation, 2.5 mi E Jamesburg (T=1, F=1); 2 mi E Jamesburg, 1700-1800 ft (T=3, M=1, F=2); 3 mi ESE Jamesburg [2000 ft] (T=8, M=3, F=5); 4 mi ESE Jamesburg, 1800-2000 ft (T=12, M=7, F=5); 5 mi ESE Jamesburg, 1800 ft (T=2, M=1, F=1); Jolon (T=7, M=5, F=2, C=1); Monterey (T=18, M=6, F=5, C=6); Pajaro (T=2, M=1, F=1, C=2); Palo Colorado Canyon, 800 ft (T=1, F=1); 1 mi E Pt. Gorda (T=1, M=1); 2 mi E Pt. Gorda (T=1, M=1, C=1); Salinas (T=2, M=1, F=1); Santa Lucia Pk., 5600 ft (T=2); Seaside [Sandune] [Live Oaks] (T=2, M=1, F=1). **San Benito County:** 7 mi SE San Juan (T=1, M=1). **Santa Cruz County:** 2.5 mi E & 1 mi N Santa Cruz, 300 ft (T=1, M=1); Santa Cruz (T=10, M=6, F=3, C=3); Freedom Blvd. near Watsonville (T=1).

SAN BENITO (11)
CALIFORNIA: Fresno County: Warthan Cr., 4.5 mi SE Priest Valley, 1850 ft (T=9, M=4, F=4). **Monterey County:** W side Arroyo Seco Wash, 150 ft, 4 mi S Soledad (T=1, M=1); Lewis Cr., 1750 ft (T=1, F=1); San Ardo, 450 ft (T=1, F=1); 0.5 mi SE San Lucas, 397 ft (T=2); San Lorenzo Cr., 1475 ft, Peachtree Valley (T=1, M=1); 1.25 mi S Soledad, 182 ft (T=1, M=1); Stonewall Cr., 1300 ft, 6.3 mi NE Soledad (T=3, M=1, F=2); Sugarloaf, 100 ft, 3 mi NNE Natividad (T=1, M=1). **San Benito County:** Cook P.O., Bear Valley (T=2, M=1, F=1); Laguna Ranch, 4000 ft, 4 mi S Hernandez (T=8, M=1, F=1); Mulberry (T=7, M=3, F=4, C=4); 2 mi NNE New Idria, 1900 ft (T=2); New Idria (T=2, M=2); Paicines (T=17, M=12, F=4, C=3); Pinnacles Nat'l Monument [Chalone Cr. Bench] (T=2, M=1, F=1); Butts Ranch, 3000 ft, 5 mi NNE San Benito (T=5); Big Oak Flat, 3300 ft, 4 mi NE San Benito (T=1, F=1); 1 mi SE summit San Benito Mtn., 4400 ft (T=4); 6 mi ESE San Benito, 1600 ft (T=9, M=1, F=2).

SAN LUIS OBISPO (12)
CALIFORNIA: Monterey County: near Bradley (T=1, M=1); Hames Valley, 1000 ft, 3 mi NE Pleyto (T=1, M=1); Parkfield (T=1, M=1). **San Luis Obispo County:** Atascadero (T=1,

M=1); Cammath [= Cammatta?] Cr., 1450 ft (T=1); S Cuyama Valley (T=1, F=1); 4 mi S Morro (T=1, F=1); Morro [= Morro Bay] (T=1, M=1); Palo Prieta Canyon [Cholame] (T=3, M=3); Paso Robles (T=12, M=7, F=5, C=9); along Salinas R., 1500 ft, 0.75 mi S Pozo (T=16, M=10, F=6, E=16); 2000 ft, Santa Lucia Mts. near San Carpojo Cr. (T=1, F=1); 1200 ft, 8 mi E Santa Margarita (T=1, F=1); Shandon (T=2, M=2).

WEST SLOPE SIERRA NEVADA-NORTH (13)

CALIFORNIA: Amador County: Carbondale (T=3, M=2, F=1); 5 mi E Carbondale (T=3, M=3); Drytown (T=2, M=1, F=1, C=1); 5 mi S Drytown (T=1, M=1); Ione (T=3, M=1); 7 mi SE Jackson (T=1, M=1, C=1); no locality (T=2, M=2, C=1). **Calaveras County:** 4 mi N Camanche (T=1, F=1). **Eldorado County:** near Cool (T=1, M=1); Greenwood (T=1, M=1); Lotus (T=1, F=1); Placerville (T=4, M=2, F=1, C=1); Middle Fork Cosumnes R., 1.7 mi SW Somerset (T=1, F=1). **Placer County:** Applegate (T=6, M=4, F=2); Clipper Gap (T=4, M=4); Colfax (T=2, M=1, F=1); 3 mi SW Colfax (T=1, F=1); Rocklin (T=1, F=1). **Sacramento County:** 2 mi NW Folsom (T=1, M=1); Folsom city (T=2, M=1).

WEST SLOPE SIERRA NEVADA-CENTRAL (14)

CALIFORNIA: Madera County: O'Neals (T=1, F=1, C=1); Raymond [940 ft] (T=7, M=5, F=1). **Mariposa County:** Dudley, 3000 ft (T=3); flat, 1850 ft, 1.75 mi W El Portal (T=1, M=1); El Portal, 2000 ft (T=11, M=4, 6); near El Portal, 3000 ft (T=1, M=1); 4000 ft, Feliciana Mtn. (T=1, F=1); Mariposa (T=1, M=1); Pleasant Valley, 600 ft (T=1). **Merced County:** Snelling (T=2, M=2). **Stanislaus County:** LaGrange (T=2, F=1); 1.5 mi SW LaGrange, 400 ft (T=3, M=2).

WEST SLOPE SIERRA NEVADA-SOUTH (15)

CALIFORNIA: Fresno County: Dunlap, 2000 ft (T=1, M=1); 3.5 mi E Dunlap, 2700 ft (T=1); 1 mi S Dunlap, 2000 ft (T=12, M=4, F=7); 3 mi W Miramonte, 2900-3500 ft (T=3, M=1, F=1); NE area Pine Flat Reservoir, 1400 ft, 0.25 mi S & 2.25 mi E Trimmer (T=15, M=12, F=3, E=15). **Tulare County:** Badger (T=2); Cottonwood Cr., 1300 ft, 0.5 mi S Aukland (T=1).

KERN (16)

CALIFORNIA: Kern County: Greenhorn Mts., 5500 ft, 8 mi W Isabella (T=2); Isabella [Sequoia Nat'l Forest] (T=6, M=1, C=1); Kelso Cr., 3750 ft, 1.1 mi E Kelso Pk. (T=1); Kern R. at Bodfish, 2400 ft (T=3, M=1); Kern R. Canyon, Greenhorn Range, Sierra Nevada mts. (T=2, M=1, F=1); Kern R. at Isabella, 2500 ft (T=4, M=1, F=1); Kern R., 5.8 air miles SW Miracle Hot Springs (T=2, M=1, F=1, C=2); Onyx [2750 ft] (T=7, M=4, F=1); Piute Mts. (T=8, M=7, F=1, C=3); Bodfish Canyon, Piute Mts. (T=4, M=3); French Gulch, 6700 ft, Piute Mts. (T=1, F=1); Rankin Ranch, 3300-3700 ft, Walker Basin (T=4, M=2, F=2); Thompson Canyon, 3900 ft, Walker Basin (T=6, M=2, F=4); E end Walker Basin, 3500 ft (T=2, F=1); Walker Pass (T=1, M=1); W slope Walker Pass [4600 ft] (T=2); 0.25 mi S & 1 mi W Walker Pass, 5400 ft, Scodie Mts. (T=1, M=1, E=1); 0.75 mi S & 1.25 mi W Walker Pass, 5800 ft, E slope Scodie Mts. (T=8, M=5, F=3, E=8); 0.75 mi S Walker Pass, 5200 ft, E slope Scodie Mts. (T=4, M=1, E=1); 1 mi S & 0.5 mi W Walker Pass, 5800 ft, Scodie Mts. (T=1, M=1, E=1); 1 mi S Walker Pass, 5800 ft, E slope Scodie Mts. (T=4, M=2, F=1, E=4); Weldon [2600-2650 ft] (T=16, M=9, F=2, C=9); Fay

Cr., 4100 ft, 6 mi N Weldon (T=6); . **Tulare County:** Hot springs, 25 mi SE Porterville (T=2, M=1, F=1); Trout Cr., 6000 ft, Sierra Nevada mts. (T=1).

SANTA BARBARA (17)

CALIFORNIA: Santa Barbara County: Buellton (T=3, M=2, F=1, C=3); De La Guerra Campground [= De La Guerra Camp] (T=1, M=1); Goleta (T=1, F=1); Goleta, 5214 Mono Dr. (T=1, M=1); near Hope Ranch (T=1); Lompoc (T=1, M=1); Los Alamos (T=2, M=2, C=1); Las Flores Ranch, Los Alamos (T=1); Nira Campground, Manzana Cr. (T=1, M=1); Old Stagecoach Hwy. (T=2, M=1, F=1, C=2); Pt. Concepcion (T=1, F=1); Santa Barbara (T=11, M=4, F=4, C=1); Mission Canyon Rd. near Botanic Garden, Santa Barbara (T=1); Rockwood Dr. near Sheffield Reservoir, Santa Barbara (T=1); Santa Ynez R. (T=2, M=1); Los Priestos Nursery, Santa Ynez Valley (T=1); San Lucas Ranch, Santa Ynez Valley (T=1, F=1); Santa Ynez Valley, near Lake Cachuma (T=1); Sunset Valley, 2300 ft, 12.5 mi NE Santa Ynez (T=1, F=1).

TEJON (18)

CALIFORNIA: Kern County: Fort Tejon (T=9, M=1, F=6, C=3); Frazier Mts., Cuddy Valley, Tecuya Mt., 6000 ft (T=1, M=1); Lebec (T=1, F=1); Mt. Pinos [6000-6500 ft] (T=5, M=4, F=1); Mt. Pinos, 5000 ft, 10 mi W Lebec (T=1, M=1). **Los Angeles County:** Angeles Nat'l Forest, Pacifico Mtn. Rd. (T=2, M=1, F=1); Liebre Mtn. (T=1, F=1); Liebre Mtn., 2 mi W Bear Camp (T=1); Liebre Mtn., 5300 ft, 1 mi S & 5 mi W Three Points (T=15, M=10, F=5, E=15); 3 mi W Three Points, 3980 ft, Liebre Mtn. region (T=6, M=3, F=2, C=6); Sandberg [4000 ft] (T=2, M=1, F=1, C=2); Sawmill Mtn., 5200+ ft, 7 mi W & 2 mi N Lake Hughes (T=1); Sawmill Mtn., USFS Rte. 7N23, vic. Sawmill Mtn. campground (T=1, M=1, C=1); Sawmill Campground, Sawmill Mtn. (T=1, F=1); Tejon Pass (T=1, M=1). **Ventura County:** 1.5 mi N & 1.5 mi W Frazier Mtn., 1720 m (T=2, E=2); Quatal Canyon (T=3, M=1, F=1, C=3).

VENTURA (19)

CALIFORNIA: Los Angeles County: Calabasas (T=1, M=1, C=1); Carbon Canyon (T=1, F=1); San Fernando Valley [= Los Angeles, San Fernando Valley] (T=9, M=5, F=4, C=6); Santa Monica Mts. (T=5, M=4, F=1); Temescal Canyon, 6 mi N Santa Monica (T=1, M=1). **Ventura County:** 5 mi N Agoura (T=3, M=2, F=1); Nordhoff (T=1, M=1); Ojai Cr. (T=1); Ojai, Ojai Valley (T=1, M=1); Ojai V. [= Valley], 1250 ft, vic. Whale Rock (T=1, F=1, C=1); Ojai Valley (T=7, M=5, F=2); Santa Paula (T=2, M=1, F=1); Santa Susana (T=3, M=1, F=2); Saticoy (T=1, F=1); 2 mi E Thousand Oaks (T=2, M=1).

NEWHALL (20)

CALIFORNIA: Los Angeles County: Bouquet Canyon (T=1, M=1); 5 mi S Lake Hughes, Elisabeth Lake Canyon (T=1, M=1); Mint Canyon, 2300 ft (T=1, M=1); Newhall (T=4, M=2, F=2); Pico Canyon, Newhall (T=1, F=1); Placerita Canyon [vic. Newhall] [35 mi N L.A. city] [1500 ft] (T=16, M=8, F=7, C=5); 5 mi E Palmdale, 2700 ft (T=2, M=1, F=1); Soledad Canyon, 14 mi NE San Fernando, 20 mi NW Mt. Wilson (T=2, M=2).

LOS ANGELES (21)

CALIFORNIA: Los Angeles County: Alhambra (T=7, M=2, F=1); Big Rock Cr., 4800 ft (T=1, M=1); Burbank (T=1, F=1); Covina (T=1, F=1, C=1); Dark Canyon (T=1, M=1); 30 mi N

Glendale, State Hwy. 2 (T=1, M=1); Glendora (T=1, M=1); near Griffith Park (T=1, F=1); Lankershim (T=2, M=1, F=1); Los Angeles (T=4, M=3, F=1, C=3); Monrovia (T=2, M=1, F=1); Mt. San Antonio (T=1, M=1); near Pasadena (T=1, F=1); Pasadena (T=34, M=20, F=13, C=7); Arroyo Seco Canyon, Pasadena, Sierra Madre mts. [= San Gabriel Mts.] (T=2, M=1); Big Santa Anita Wash, Pasadena (T=1, M=1); Eaton's Wash, near Pasadena (T=3, M=2, F=1); Eaton Canyon (T=1, M=1); Millard's Canyon, Pasadena (T=2, M=1, F=1); Millard Canyon, Altadena (T=3, M=2, F=1, C=3); San Dimas Canyon (T=7; M=5, F=2, C=2); San Gabriel (T=1, M=1); Barley Flats, San Gabriel Mts. [= Sierra Madre mts.] (T=4, M=3, F=1, C=3); Chilao Flats [San Gabriel Mts.] (T=1, M=1, C=1); Heninger Flats, 3000 ft, San Gabriel Mts. (T=1, M=1); Horse Flats, 5600 ft, Angeles Nat'l Forest, San Gabriel Mts. (T=2, M=2); W Fork San Gabriel R., 2000 ft, 18 mi above Azusa, Sierra Madre mts. [= San Gabriel Mts.] (T=1, F=1); Pine Flats (T=1, M=1); Santa Anita (T=1, F=1); Oak Crest [Cahuenga Pass] [Santa Monica Mts.] (T=14, M=5, F=7, C=7); near Sierra Madre (T=1, C=1); Wilson's Pk. (T=1); Mt. Wilson [Sierra Madre = San Gabriel Mts.][5600 ft] (T=5, M=3, F=2, C=1). **San Bernardino County:** Carbon Canyon, 4 mi NE county line (T=1, M=1); Red Hill near Upland, 1200 ft (T=1, M=1).

SAN BERNARDINO MTS. (22)
CALIFORNIA: Riverside County: near Beaumont (T=2, M=1); 2 mi S Redlands, 1500 ft (T=1, M=1); Riverside (T=3, M=1, F=2). **San Bernardino County:** Cajon Pass (T=2, M=1, C=1); E of Lone Pine, near Cajon (T=1, F=1); 7 mi N Crestline, Miller Canyon, 5000 ft (T=1, F=1); Redlands (T=9, M=6, F=3); 1600 ft, 2 mi S Redlands (T=1, F=1); 4 mi S Redlands, 1800 ft (T=1); San Bernardino (T=1, M=1); San Bernardino Mts. (T=2, M=1, F=1, C=2); San Bernardino Mts., the Pipes Canyon, 6000 ft, 35 mi W Twentynine Palms (T=1, M=1); Cactus Flat, 6000 ft, San Bernardino Mts. (T=2); Oak Glen [5300 ft, San Bernardino Mts.] (T=11, M=9, F=1, C=9); Whiteside Canyon, San Bernardino Mts. (T=1, M=1); San Gabriel Mts., N Fork Lytle Cr. Canyon, 4000 ft (T=2, M=1, F=1); 10 mi N San Gorgonio (T=2, M=2); 7 mi NW Yucaipa (T=1, F=1); Yucaipa (T=1, F=1); Hog Canyon, Yucaipa (T=2, M=1, F=1); Wildwood Park, Yucaipa (T=1).

SAN JACINTO MTS. (23)
CALIFORNIA: Riverside County: Palm Springs (T=1, F=1); San Jacinto Mts. (T=7, M=6, C=2); Kenworthy, 4500 ft, San Jacinto Mts. (T=1, M=1); Palm Canyon, 3000 ft, San Jacinto Mts. (T=1, F=1); Poppet Flat, 3700 ft, San Jacinto Mts. (T=2); Schain's Ranch, 4960 ft, San Jacinto Mts. (T=1); Strawberry Valley, 6000 ft, San Jacinto Mts. (T=3, M=1); Santa Rosa Mts., 4 mi SW Spring Crest (T=1, M=1); 3 mi NW Rabbit Pk., 3500-4200 ft [Santa Rosa Mts.] (T=6, M=1, F=4); Garnet Queen Mine, 6000 ft, Santa Rosa Mts. (T=4, M=1, F=1); Whitewater R. delta near Mecca (T=1, F=1).

JOSHUA TREE (24)
CALIFORNIA: Riverside County: Joshua Tree Nat'l Monument, lower Covington Flats (T=1, M=1); lower Covington Flat, 5000 ft (T=2, M=1, F=1); Joshua Tree Nat'l Monument, upper Covington Flats (T=1, M=1); Pinyon Wells, 4000 ft, Joshua Tree Nat'l Monument (T=1, M=1); Pinyon Wells, 4000-4500 ft (T=9, M=5, F=4); Pinyon Well, 16.25 mi S & 2.5 mi W Twentynine Palms, 4000 ft (T=1, M=1); Quail Spring, Little San Bernardino Mts. (T=2, M=1, F=1); Joshua Tree Nat'l Monument, Smithwater Canyon (T=1, M=1); Joshua Tree Nat'l Monument, Smithwater

Wash (T=3, M=2, F=1); Juniper Flat, 4800 ft, 2 mi N & 1 mi E Stubby Spring, Little San Bernardino Mts., Joshua Tree Nat'l Monument (T=5, M=3, E=5); Juniper Flat, 4750 ft, 1.25 mi N & 0.5 mi E Stubby Spring, Little San Bernardino Mts., Joshua Tree Nat'l Monument (T=1, M=1, E=1); 2 mi N Stubby Spring, 4700 ft, Little San Bernardino Mts. (T=1, F=1, C=1); Stubby Spring, Joshua Tree Nat'l Monument (T=1, M=1, C=1); Stubby Spring[s] [4500-4750 ft] [Little San Bernardino Mts.] (T=10, M=6, F=4, C=5); 1.25 mi E Stubby Spring, 5000 ft, Little San Bernardino Mts. (T=1, F=1, E=1). **San Bernardino County:** Black Rock Canyon, 4800 ft, Little San Bernardino Mts. (T=1, M=1, E=1); Black Rock Canyon, 4600 ft, 3.75 mi S & 2 mi E Burnt Mtn., Little San Bernardino Mts. (T=2, M=1, F=1, E=2); Black Rock Spring, 4500 ft (T=15, M=6, F=6, C=1); Quail Spring, 3600-4500 ft [Joshua Tree Nat'l Monument] (T=6, M=4, F=2).

ORANGE (25)

CALIFORNIA: Orange County: Capistrano (T=1, F=1); Eucalyptus grove, 7th St., Los Alamitos Blvd. (T=1, M=1); Hwy. 74, Cleveland Nat'l Forest (T=1, F=1); lower San Juan Camp (T=1, M=1); Santa Ana Canyon, 400 ft (T=1, M=1); Trabuco Canyon, 1700 ft (T=1). **Riverside County:** Elsinore (T=2, M=2). **San Diego County:** Escondido (T=4, M=2, F=1); Mesa Grande (T=1, M=1); Palomar Mtn. (T=1, M=1); 2.7 mi W & 1.3 mi N Rainbow, 500 ft (T=1); San Marcos (T=3, M=1, F=2, C=1).

SAN DIEGO-NORTH (26)

CALIFORNIA: San Diego County: Ballena (T=6, M=4); Banner Grade, 3200 ft (T=1, M=1); 2.85 mi NE Cuyamaca Lodge, Hwy. 79 (T=1, F=1); Eagle Pk. (T=1, F=1); 2 mi E El Monte Park (T=1, F=1); Flinn Springs (T=1, M=1); Julian (T=5, M=4, F=1); La Puerta Valley (T=1, F=1); Monte Robles, 4 mi SW Ramona (T=1, F=1); Pine Hills (T=1, M=1); Santa Ysabel (T=2, M=2); Volcan Mtn. [Mts.] (T=3, M=2, F=1); 10 mi SE Warner Springs (T=1, M=1); Witch Cr. (T=60, M=28, F=30, C=16).

SAN DIEGO-SOUTH (27)

MEXICO: Baja California Norte: [N end] Nachoguero Valley [3400 ft] (T=27, M=15, F=12); Nachoguero Valley (T=8, M=4, F=4).

CALIFORNIA: San Diego County: Campo (T=6, M=4); near Campo (T=3, M=2); Cibbets Flat (T=1, M=1, C=1); Cuyamaca Mts. (T=5, M=2, F=2); between Deerhorn Flat & Honey Springs Ranch (T=1, F=1); 8.5 mi E & 0.5 mi S Descanso, 5100 ft (T=1); Dulzura (T=19, M=14, F=5, C=1); Guatay, Cleveland Nat'l Forest (T=1, F=1, C=1); 4 mi E Guatay (T=2, M=1, F=1); Horsethief Canyon, 2200 ft, 6.75 mi S & 2.5 mi W Descanso (T=10, M=6, F=4, E=10); Horsethief Canyon, 6 mi SSW Descanso, 2400 ft (T=1, M=1); 1 mi E Jcn. 79 & 80 (along 80)- fire access rd. (T=2); Kitchen Cr., 1 mi N Cibbetts Flat (T=1, F=1, C=1); Kitchen Cr., [4200 ft], above Cibbetts Flat, Laguna Mt. [Mts.] (T=3, M=1, F=1, E=2); Laguna (T=2, M=1); Campbell's Ranch, Laguna Mts. (T=2); Morris Ranch, Laguna Mt., 5500 ft (T=2, M=1, F=1); 1.3 mi SW Wooded Hill, Laguna Mts. (T=1, M=1); NE Morena Conservation Camp, 3200 ft, 2.5 mi N Morena Village (T=5, M=4, F=1, E=5); Morena Lake (T=1, F=1); 2.5 mi NE Pine Valley (T=1, M=1); 2 mi E Pine Valley (T=2, M=2); 1.5 mi SW Pine Valley (T=2, M=1, F=1); Rancho Corte Madera (T=7, M=3, F=3); Sweetwater R. (T=1, M=1); Thing Valley (T=3, M=1, F=2); Vallecito Stage Station, 1500 ft (T=1, F=1).

SIERRA JUAREZ (28)

MEXICO: Baja California Norte: Burro Canyon [25 mi NE Ensenada] (T=3, M=1, F=2); Las Cruces, 2600 ft, 20 mi E Ensenada (T=6, M=3, F=2); El Rayo, Sierra Juarez (T=2, F=2); El Rayo, Hansen Laguna Mts. (T=1, M=1); Laguna Hansen, Sierra Juarez [5200 ft] (T=20, M=10, F=7, C=11); Los Pozos, 4200 ft, Sierra Juarez (T=2, M=1, F=1).

SIERRA SAN PEDRO MÁRTIR (29)

MEXICO: Baja California Norte: 10 mi SE Alamo [Rancho San Pablo] (T=13, M=6, F=7, C=11); 17 mi E Rancho San José de Castillo (T=1, F=1); W slope San Pedro [= Sierra San Pedro Mártir] (T=1, M=1); San Pedro Mts. (T=2, F=1); Concepcion, 6000 ft, San Pedro Mártir Mts. (T=8, M=4, F=4); La Joya, 6200 ft, San Pedro Mártir Mts. (T=1, M=1); 2 mi S La Joya, Valladares Cr., San Pedro Mártir Mts. (T=3); San Antonio, W base San Pedro Mártir Mts. (T=1, F=1); Santo Domingo R., 3 mi N San Antonio (T=1); San Antonio Ranch, 2100 ft, Santo Domingo R. (T=1, M=1); El Valle de la Trinidad, 2500 ft [= Valle Trinidad, 2500 ft] (T=6, M=5, F=1).

SIERRA DE LA LAGUNA (30)

MEXICO: Baja California Sur: Mt. Miraflores (T=6, M=2, F=4, C=1); summit Sierra de la Laguna (T=1); Sierra Laguna [= Sierra de la Laguna mts. = Sierra de la Laguna] [7200 ft] (T=94, M=44, F=34, C=13); El Sauz, Sierra Laguna (T=2, M=1, F=1, C=2); La Chuparosa, Sierra Laguna (T=1, M=1); La Laguna, Sierra Laguna [5600 ft] (T=16, M=8, F=5, C=7); Victoria Mts. (T=3, F=1); on ridge, 5700 ft, 3 mi SW El Sance, Victoria Mts. (T=2, M=1, F=1, C=2); Laguna Valley [6000 ft, Victoria Mts.] (T=52, M=27, F=17).

WARNER (31)

NEVADA: Washoe County: Mud Lake (T=2, M=1, F=1).

OREGON: Lake County: Jacob's Ranch, 8 mi S Adel (T=4, M=2, F=2); W rim Twentymile Cr. Canyon, 5300 ft, 8 mi S & 4 mi W Adel (T=21, M=8, F=7, E=16); 9 mi S Adel, W rim [=ridge] Twentymile Cr. [Canyon] (T=4, M=4); 9 mi S Adel, mouth of Twentymile Cr. (T=1, M=1); hills at Twentymile Cr. Canyon, 9 mi S Adel (T=1, F=1); hills at mouth of Twentymile Cr., 9 mi S Adel (T=1, M=1); S Warner Valley, 8 mi S Adel (T=2, M=1, F=1).

TULE LAKE (32)

CALIFORNIA: Modoc County: Clear Lake (T=2, M=1, F=1); Steele Meadow, 4700 ft (T=2, M=1, F=1); **near Tule Lake Hwy. (Hwy. 139), 4250 ft, 1.25 mi N & 3 mi E Timber Mtn.** (T=13, M=8, F=4, E=13).

OREGON: Klamath County: 6 mi E Lorella (T=2, M=1, F=1).

NORTHEASTERN CALIFORNIA (33)

CALIFORNIA: Lassen County: Juniper Ridge, 5600 ft, 2.5 mi S & 2.5 mi E McDonald Pk. (T=3, M=2, F=1, E=3); Juniper Ridge, 5600 ft, 3.5 mi S & 3 mi E McDonald Pk. (T=11, M=4, F=2, E=7); 2 mi S & 3 mi E McDonald Pk., 5700 ft (T=11, M=6, F=2, E=8); 4 mi SW McDonald Pk., 5300 ft (T=12, M=7, F=5, C=12); 8 mi SW Ravendale, 5600 ft (T=8, M=3, F=2); 2 mi N Red Rock P.O., 5300 ft (T=1, M=1); Secret Valley, 5000 ft (T=8, M=3, F=3); Shinns Pk.,

5100 ft, 10 mi S & 3 mi E Ravendale (T=2, M=1). **Modoc County:** Devil's Garden (T=2, M=1, F=1); 3 mi N Eagleville (T=1); Eagleville (T=1, M=1); Juniper Scott Ranch, 5000 ft, 10 mi SW Alturas (T=1, F=1).

RENO (34)

CALIFORNIA: Lassen County: SW slope Red Rock Canyon, 4700 ft, 0.5 mi N & 0.5 mi W Red Rock Pk. (T=4, M=2, F=1, E=4).

NEVADA: Douglas County: Minden (T=1, F=1); Pine Nut Mts., U.S. 395, 1.4 mi SE Palamino Rd. (T=1, F=1); Pine Nut Range, 6000 ft, 8.5 mi S & 7 mi E Gardnerville (T=1, M=1). **Lyon County:** Como District Rd. (T=1, F=1). **Ormsby [= Carson City] County:** Brunswick Canyon (T=1); Sand Canyon, near Brunswick Canyon (T=1). **Storey County:** Virginia City [6500 ft] (T=2, M=2). **Washoe County:** S slope Dry Valley Cr., 4900 ft, 2 mi N & 5 mi E Red Rock Pk. (T=5, M=3, F=2, E=5); S slope Red Rock Canyon, 5000 ft, 0.75 mi N & 0.25 mi W Red Rock Pk. (T=6, M=2, F=2, E=4); 17 mi NW Reno, 6000 ft (T=12, M=6, F=6); Reno (T=4, M=2, F=2, C=1); 5200 ft, [W] foothills Virginia Range, 2 mi E Steamboat (T=2, M=2); 2 mi E & 1 mi S Steamboat Springs (T=2, M=2); 6.5 mi ESE Tule Pk., 4500 ft, Virginia Mts. (T=1, F=1).

BENTON (35)

CALIFORNIA: Mono County: W base Anchorite Hills, 7000 ft, along Calif.-Nevada state line, 2.25 mi S & 5.75 mi W Anchorite Pass (T=1, F=1, E=1); W slope Benton Range, 6800 ft, 4 mi N & 5 mi W Benton (T=6, M=1, F=1, E=2); 5 mi W & 4 mi N Benton, 6800 ft (T=3, M=2, F=1); 6 mi W & 3 mi N Benton, 7000 ft (T=2, M=2); 8 mi W & 2 mi N Benton, 7500 ft (T=1, F=1); Benton, 5639 ft (T=1. F=1); 3.5 mi W & 3.5 mi S Benton, 7000 ft (T=1); W slope Benton Range, 6900 ft, 4 mi S & 6 mi W Benton (T=1, M=1, E=1).

NEVADA: Mineral County: Anchorite Hills, 7100 ft, 3 mi S & 3 mi W Anchorite Pass (T=3, M=2, F=1, E=3); W base Anchorite Hills, 7000 ft, 2.25 mi S & 5 mi W Anchorite Pass (T=7, M=3, F=3, E=7); W base Anchorite Hills, 7000 ft, along Calif.-Nevada state line, 2.25 mi S & 5.75 mi W Anchorite Pass (T=1, F=1, E=1); Powell Canyon, 7500 ft, 1 mi E Powell Mtn., Wassuk Range (T=1, F=1).

WHITE-INYO MOUNTAINS (36)

CALIFORNIA: Inyo County: Cedar Flat, 7400 ft, 0.25 mi W Westgard Pass, White-Inyo Mts. (T=3, M=1, F=1, E=2); Cedar Flat, 7100 ft, 1 mi S & 0.5 mi E Westgard Pass, White-Inyo Mts. (T=3, M=2, F=1, E=3); Cedar Flat, 7100 ft, 1.25 mi S Westgard Pass, White-Inyo Mts. (T=1, M=1, E=1); Cedar Flat, 7200 ft, 1.75 mi S & 0.5 mi E Westgard Pass, White-Inyo Mts. (T=2, M=1, F=1, E=2); Cedar Flat, 7200 ft, 2 mi S & 0.5 mi E Westgard Pass, White-Inyo Mts. (T=9, M=5, F=1, E=7); Cedar Flat, 7300-7500 ft, White Mts. (T=10, M=3, F=7, C=7); Westgard Pass, 7300 ft (T=6, M=3, F=3, C=5); Westgard Pass, White Mts. (T=2, M=1, F=1); near summit Westgard Pass (T=4); White Mts., 7500 ft (T=17, M=8, F=7); E base Waucoba Mtn., 7300 ft (T=9, M=6, F=3); Waucoba Pass, 7500 ft, Inyo Mts. (T=4, M=2, F=2).

EASTERN MOJAVE (37)

CALIFORNIA: San Bernardino County: Clark Mtn., 6000-7000 ft [E San Bernardino Co.] (T=3, M=2, F=1); Colosseum Mine, Clark Mtn. (T=2); SE side Clark Mtn. [5800-6300 ft] (T=9,

M=8, F=1); Cottonwood Spring, 4400 ft, Granite Mts. (T=1, M=1); [E Mojave Desert] New York Mts., Carothers Canyon, 5640 ft (T=10, M=2, F=1, C=3, E=9); New York Mts. (T=1, M=1); New York Mts., 5 mi S Ivanpah, 5000 ft (T=2, M=1, F=1); Bonanza King Canyon, Providence Mts. (T=1, F=1); Cedar Canyon, 5000-5800 ft, Providence Mts. (T=49, M=26, F=20, E=6); 6 mi E & 3 mi S Cima, 5500 ft, Providence Mts. (T=1, F=1); 3 mi S & 7 mi E Cima (T=3, F=1); Midhills, 5 mi E & 4 mi S Cima (T=1, F=1); 12 mi NE Granite Well, 5400 ft, Providence Mts. (T=1, F=1); 5 mi NE Granite Well, 5400 ft, Providence Mts. (T=13, M=6, F=6); 5 mi E Kelso in Wildhorse Canyon, Providence Mts. (T=2, M=1, F=1); Mitchell's Caverns, 4500 ft, Providence Mts. (T=1, F=1).

SOUTHERN NEVADA (38)

NEVADA: Clark County: Charleston Mts., 6200-6700 ft (T=2, M=2, C=1); N side Potosi Mtn., 6000-7000 ft (T=5, M=1, F=1); Sheep Mts., Mormon Well (T=1, F=1); Hidden Forest Canyon, 6400-6900 ft, Sheep Range (T=4, F=1); 0.5 mi E Macks Canyon, 7700 ft, Spring Mts. (T=2). **Nye County:** Mercury, Nevada Test Site (T=2, M=1).

SOUTHEASTERN NEVADA (39)

NEVADA: Lincoln County: Ash Canyon, 5700 ft, 4 mi N & 1 mi E Ella Mtn., Clover Mts. (T=7, M=2, F=1, E=3); 1.5 mi N & 2.5 mi E Ella Mtn., 6000 ft, Clover Mts. (T=15, M=5, F=3, E=8); 1 mi W Panaca Summit, 6200 ft, Cedar Range (T=1, M=1); 1 mi S & 9 mi W Panaca Summit, 6300 ft, Cedar Range (T=3, M=2, F=1, E=3); **0.5 mi S Oak Springs Summit, 6300 ft, Delamar Range** (T=1, M=1, E=1); Springer Spring, 7000 ft, Mt. Irish (T=4, M=2, F=2); **NE slope Timber Mtn., 6100-6600 ft, 3.25 mi N & 0.75 mi E Weepah Spring, Seaman Range** (T=3, M=2, F=1, E=3). **Nye County:** Garden Valley, 8.5 mi NE Sharp (T=1, F=1).

RUBY MOUNTAINS (40)

NEVADA: Elko County: W side Ruby Lake, 6 mi N Elko Co. line, 7000 ft (T=2, M=1, F=1). **White Pine County:** W side Ruby Lake, 3 mi S White Pine Co. line, 6100 ft (T=3, M=2, F=1); 0.5 mi S & 1.5 mi E Overland Pass, 6600 ft, S end Ruby Mts. (T=6, M=3, F=2, E=4); 0.5 mi S & 3 mi E Overland Pass, 6600 ft, S end Ruby Mts. (T=4, M=4, E=4); 0.75 mi S & 4.5 mi E Overland Pass, 6500 ft, S end Ruby Mts. (T=4, M=2, F=2, E=4); E slope Ruby Mts., 6100 ft, 0.25 mi S & 3 mi E Sherman Mtn. (T=3, M=2, F=1, E=3); Willow Cr., 2 mi S White Pine Co. line, Ruby Mts. (T=1, M=1).

SOUTHERN IDAHO (41)

IDAHO: Bannock County: Pocatello (T=8, M=3, F=5, C=4); Mink Cr., 12 mi S Pocatello (T=1, M=1); Pocatello Cr., 3 mi E Pocatello (T=2, M=1, F=1); 5 mi SE Pocatello (T=1, F=1). **Cassia County:** Bridge (T=2, M=1, F=1); Elba (T=3, M=2, F=1); Oakley [Goose Cr.] (T=6, M=2, F=2, C=4); 10 mi S Oakley, Goose Cr. (T=2, M=1, F=1). **Minidoka County:** Minidoka (T=1, F=1).

UTAH: Box Elder County: 5 mi N Grouse Cr., 5324-5400 ft (T=4, M=1); 1 mi N Grouse Cr., 5300 ft (T=1); Grouse Cr., 5300 ft (T=2, F=1); Clear Cr., 6500 ft, N slope Raft River Mts., 5 mi SW Rafton (T=2, M=1, F=1); 1 mi N Rosette, 6000 ft, S slope Raft River Mts. (T=2, M=2); Standrod, 10 mi W NAF Idaho, 98 mi W Brigham Utah at One Mile Cr. (T=1, F=1); 3 mi SE Yost, 6000 ft (T=1, M=1).

NORTH-CENTRAL UTAH (42)

UTAH: Juab County: Death Canyon, S [SW] end Simpson Mts. [4800 ft] (T=2, M=1, F=1, C=2); 12 mi NW Jericho, 4400 ft (T=8, M=5, F=3). **Tooele County:** Black Sage Pass, S end Cedar Mts., 4700 ft (T=2, M=1, F=1, C=2); Cane Springs, 5400 ft, 4.5 mi E Wig Mtn., Cedar Mts. (T=2); Sandy Pass, SW end Cedar Mts., 4600-4700 ft (T=2, M=1, F=1, C=2); N foothills Little Davis Mtn. [4800 ft] (T=2, F=1); Cane Spring, W side Stansbury Mts., 5000 ft (T=1); Callister's Ranch, 5600 ft, Box Canyon, 4 mi NE Iosepa (Deseret Livestock Ranch), NW end Stansbury Mts. (T=2, M=1, F=1); W side Stansbury Mts., W Hickman Canyon, 5700 ft, 6 mi N Willow Spring (T=1, F=1); 3 mi W Willow Springs, 4500 ft, W side Stansbury Mts. (T=1, M=1); near Willow Springs, 4600 ft, W side Johnson Pass, Stansbury Mts. (T=3, M=3, C=2); near Johnson Pass, 6500 ft, 2 mi E Willow Spring, S end Stansbury Mts. (T=16, M=6, F=3, E=10); Willow Spring, Skull Valley (T=1, F=1); 1 mi N & 1.5 mi E Johnson Pass, 6100 ft, SE end Stansbury Mts. (T=8, M=2, F=1, E=4); 0.5 mi S & 2.25 mi E Johnson Pass, 5700 ft, SE end Stansbury Mts. (T=5, M=1, F=2, E=5); Clover Cr., near Johnson's Pass, 5950 ft (T=1, C=1); Rock Springs Canyon, Stansbury Mts., 5500 ft (T=1, M=1); W side Lookout Pass, Onaqui Mts., 6000 ft (T=1, F=1, C=1); Jurgenson's Ranch, mouth of Ophir Canyon, 5000 ft (T=1, M=1); Lofgren Spring, S end Rush Valley, 6000 ft (T=1). **Utah County:** 1 mi N Cedar Fort, 4600 ft (T=1, M=1).

CENTRAL UTAH (43)

UTAH: Beaver County: Beaver (T=3, M=2, F=1, C=2). **Emery County:** Huntington Canyon, 6 mi NW Huntington, 6100 ft (T=2, F=1, C=2). **Garfield County:** 5 mi N Boulder, 7500 ft, SE base Aquarius Plateau (T=2, M=1, F=1); Jcn. Calf Cr. & Escalante R., 5200 ft, 10 mi E Escalante (T=1, F=1); 8 mi S Escalante, 5200 ft (T=2, M=1); Sanford Ranch, 6000 ft, W slope Mt. Hiller, Henry Mts. (T=1, M=1); Sanford's Ranch, 6000 ft, 30 mi S Hanksville, N base Mt. Hiller, Henry Mts. (T=1, M=1). **Sanpete County:** 4 mi S Spring City (T=1, M=1). **Sevier County:** 5 mi SE Salina, 6000 ft (T=1, F=1). **Wayne County:** 7300 ft near Grover, NE foothills Aquarius Plateau (T=1, M=1).

LA SAL (44)

COLORADO: Montrose County: W Paradox Valley (T=1, M=1).

UTAH: Grand County: Mill Cr. Canyon, 5000 ft, 4 mi NE Moab (T=1, C=1); 15 mi SE Moab in La Salle [=La Sal] Mts., 6000 ft (T=2). **San Juan County:** Big Indian Wash, 34 mi SE Moab, 6000 ft (T=1, M=1); Black Canyon, 5400 ft, near Kane Springs, 19 mi S [SE] Moab (T=17, M=7, F=10); Browns Hole, Black Canyon, 25 mi SE Moab, 6000 ft (T=1, F=1); Muleshoe, Black Canyon, 3 mi S Cane [=Kane] Spring, 5500 ft (T=1, M=1); Muleshoe camp, Black Canyon, near Cane [=Kane] Spring, 19 mi SE Moab (T=1, F=1); stock tank, near Muleshoe, 20 mi SE Moab, 5500 ft (T=1, F=1); W slope La Sal Mts., 6400 ft, 2.5 mi N & 3.5 mi E La Sal Jcn. (T=4, M=3, F=1, E=4); W slope La Sal Mts., 5800-6200 ft, 2.25 mi N & 0.75 mi E La Sal Jcn. (T=12, M=4, F=6, E=12).

NEW CASTLE (45)

COLORADO: Garfield County: Glenwood Springs (T=1); New Castle (T=26, M=13, F=11, C=3); 6200 ft, 7 mi S & 3 mi W New Castle (T=5, M=1, F=3, E=5); 6600 ft, 8 mi S &

2.25 mi W New Castle (T=11, M=4, F=3, E=11). **Rio Blanco County:** L07 Gulch (T=1, C=1); Windy Gulch (T=5, F=1, C=3).

PUEBLO (46)

COLORADO: Chaffee County: Buena Vista (T=1, F=1); Salida (T=1, F=1). **El Paso County:** no locality (T=3, M=1, F=1). **Fremont County:** 7 mi W Canon City (T=1, M=1); Canon City (T=2, M=1, F=1, C=1); Garden Park (T=2, F=1, C=1). **Pueblo County:** no locality (T=4, M=2, F=2).

SOUTHWESTERN PLAINS (47)

COLORADO: Bent County: Ninaview (T=1, M=1). **Las Animas County:** Sopris (T=1, M=1); Trinidad (T=11, M=3, F=5, C=1); near Trinidad (T=1, M=1); 7 mi E & 1 mi S Trinidad (T=1); Watervale (T=2).

NEW MEXICO: Harding County: Mills Canyon, ca. 10 mi W Mills on Canadian R. (T=1, M=1). **Union County:** 7500 ft, rimrock of Mesa, Folsom (T=1, M=1).

OKLAHOMA: Cimarron County: Kenton (T=5, M=2, F=2, C=4); 1 mi S Kenton (T=1, M=1); 1 mi SE Kenton (T=1, M=1); 4 mi S Kenton (T=2, M=1, F=1); 4 mi SE Kenton (T=1, M=1); 5 mi S Kenton (T=1, M=1); 6 mi S Kenton (T=1, M=1); 6 mi SE Kenton (T=7, M=6, F=1).

NORTH-CENTRAL NEW MEXICO (48)

NEW MEXICO: Bernalillo County: Cedar Crest (T=1, M=1); Manzano Mts. (T=1, M=1); Carolina Canyon, Manzano Mts. (T=1); 2 mi S Tijeras (T=1, M=1); 2 mi E & 2.5 mi S Tijeras, 7180 ft (T=1, C=1); 3.5 mi E & 3.5 mi S Tijeras, 7460 ft (T=2, C=2); 8 mi S Tijeras (T=1, M=1); Tijeras Canyon, Sandia Mts. (T=1, M=1). **Rio Arriba County:** 5 mi N El Rito, 7000-7100 ft (T=2, M=1); 4.5 mi NW El Rito, 7500 ft (T=1); 4 mi N El Rito, 7000 ft (T=8); Gallina (T=1, F=1). **Sandoval County:** 2 mi N & 1 mi W San Luis (T=1, M=1). **San Miguel County:** Rowe (T=2, M=1, F=1); Santa Fe Nat'l Forest above Pecos (T=2, M=1, F=1, C=2). **Santa Fe County:** San Pedro (T=1, F=1); Santa Fe (T=1). **Taos County:** 7000 ft, 3 mi N & 6 mi E Ojo Caliente (T=20, M=10, F=6, E=20); Questa (T=2, M=1). **Torrance County:** 2 mi S Escabosa (T=1).

ZUNI MOUNTAINS (49)

NEW MEXICO: McKinley County: Horsehead Canyon, 1 mi S & 5.5 mi E Blackrock, 6900 ft (T=3, M=2, F=1, E=3); Sixmile Canyon, 7100 ft, 0.5 mi S & 4.5 mi E Fort Wingate, N end Zuni Mts. (T=14, M=7, F=6, E=14).

FOUR CORNERS (50)

ARIZONA: Apache County: NW base Carrizo Mts. (T=1, M=1, C=1). **Navajo County:** Bubbling Springs Canyon, 6500 ft (T=1); mouth of Long Canyon, 6700 ft (T=1); 4 mi NE Marsh Pass camp, 8 mi W Kayenta, 6000 ft (T=1, F=1); 5 mi SW Marsh Pass camp at mouth of Tsegi, 6500 ft (T=1, F=1); Tsegi Canyon, 6000 ft, 12.2 mi SWW Kayenta (T=1, M=1).

COLORADO: La Plata County: near Bondad (T=4, M=2, F=1). **Montezuma County:** Mesa Verde (T=1, F=1); Ute Pk. (T=1, M=1).

NEW MEXICO: San Juan County: 3 mi E Archuleta, 5800 ft (T=1); 5 mi E Archuleta (T=2,

M=1, F=1); Aztec (T=1); opposite mouth of Bancos Canyon, river mile 163, San Juan R. (T=2); Rio San Juan (T=2, M=1).

SOUTHWESTERN UTAH (51)

UTAH: Iron County: 10 mi W Cedar City (T=1, F=1); Iron City (T=1, F=1); 6000 ft, Iron City ruins (T=1, F=1, C=1); 2 mi SE Irontown ruins, 6500 ft (T=3). **Washington County:** Beaver Dam Wash, 2250 ft (T=1, M=1); 3.5 mi S Enterprise (T=1, M=1); Jackson Spring (T=1); Danish Ranch, 4200 ft, 5 mi NW Leeds (T=4, M=3, F=1); 2 mi N Pinto, 6600 ft (T=4, M=2, F=1); 2 mi SE Pinto, 6700 ft (T=2, M=1, F=1); Pintura, 3500 ft (T=1, M=1, C=1); 6 mi N Springdale, 4300 ft (T=1, M=1); 6500 ft [no locality] (T=1, F=1).

KAIBAB (52)

ARIZONA: Coconino County: Le Fevre Canyon, 6100 ft, Kaibab Natl Forest, 5.5 mi S & 15 mi E Fredonia (T=4, M=3, F=1, E=4); Le Fevre Canyon, 6500 ft, Kaibab Natl Forest, 7 mi S & 15 mi E Fredonia (T=10, M=6, F=4, E=10); W slope Le Fevre Ridge, 6500 ft, Kaibab Natl Forest, 7 mi S & 14.5 mi E Fredonia (T=1, M=1, E=1). **Mojave County:** Nixon Spring, Mt. Trumbull (T=2); 6250 ft, Nixon Spring, Mt. Trumbull (T=1, M=1).

UTAH: Kane County: Hamblin [?] Ranch, 5500 ft, Cave Lakes Canyon, 5 mi NW Kanab (T=2, M=1, F=1); Red Canyon, 5700 ft, 6 mi N Kanab (T=4, M=1, F=3, C=3).

SAN FRANCISCO MOUNTAINS (53)

ARIZONA: Coconino County: 2.5 mi E Anita (T=1, M=1); 2 mi NNE Coconino, 6500 ft, Kaibab Plateau (T=1); Deadman's Flat, San Francisco Mts. (T=1); Deadman Flat, 5500 ft, San Francisco Mts. (T=1, F=1); Upper Deadman's Flat, San Francisco Mts. (T=1, M=1); Deadman's Flat, 6400-7000 ft, NE San Francisco Mt. (T=16, M=8, F=8); Deadman Flat, 6200 ft, NE San Francisco Mt., 20 mi N & 6.5 mi E Flagstaff (T=10, M=3, F=4, E=10); Deadman Mesa, 6400 ft, NE base San Francisco Mt., 17 mi N & 6.5 mi E Flagstaff (T=7, M=2, F=4, E=7); Doney Mt., San Francisco Mts. (T=2, F=1); San Francisco Mt., 6000 ft (T=3, F=1, C=1); San Francisco Mts. (T=1, M=1, C=1); 6550 ft, tank on Old Grand Canyon Stage Rd., 38 mi N Flagstaff (T=1, F=1); 10 mi E Flagstaff (T=1, M=1); N of Williams [T22N, R2E, Section 3] (T=2, M=1); Williams (T=1, F=1). **Yavapai County:** Ash Fork (T=2, M=1, F=1).

CENTRAL ARIZONA (54)

ARIZONA: Coconino County: 5 mi S Sedona, 4200 ft (T=1, M=1, C=1); 5 mi S Stoneman Lake (T=1, F=1). **Yavapai County:** near Fort [= Camp] Verde (T=2, M=2); Fort [= Camp] Verde (T=1); Montezuma Well, 3500 ft (T=1, M=1); Payson (T=3, M=1, F=2); 10 mi W Prescott (T=1, M=1, C=1); Prescott (T=4, M=3); Prescott, Walker Rd., 0.8 mi S jct. Hwy. 69 (T=3, M=2, F=1, C=2); Aspen Cr., Prescott (T=1, M=1, C=1); Dells, Prescott (T=2, M=1, F=1); Dry Cr., 7 mi SW Sedona (T=2, M=1, F=1); Fort Whipple [= Whipple] (T=1, M=1, C-1); Fort Whipple, Prescott (T=1, F=1); 5 mi E Whipple Barracks (T=1, C=1); Williamson Valley, Prescott (T=2, M=1, F=1).

EASTERN ARIZONA (55)

ARIZONA: Gila County: Sawmill, 27 mi NE Globe, 5600 ft (T=1, M=1). **Navajo County:** Fort Apache (T=1); Show Low Cr. (T=11, M=8, F=3); Taylor (T=16, M=9, F=7, C=7); 8 mi S White R., 6100 ft (T=8, M=6, F=2).

RESERVE (56)

NEW MEXICO: Catron County: Frisco (T=1, F=1); 1 mi N & 2.5 mi E Reserve, 6100 ft (T=8, M=5, F=2, E=7); Reserve [Saliz Canyon] (T=106, M=57, F=45, C=5); 6 mi S Reserve, 6600 ft (T=7, M=4, F=3, E=7); Gatlin Canyon, 6500 ft, 7 mi S Reserve (T=1, M=1, E=1).

SILVER CITY (57)

NEW MEXICO: Grant County: Burro Mts. (T=1); Fierro (T=3, M=3); Fort Bayard (T=2, M=2); Gila R. at mouth of Mogollon Cr. (T=1); 25 mi NE Lordsburg (T=5, M=4, F=1, C=5); Lone Mtn. [near Silver City] (T=2, M=2); Silver City (T=13, M=6, F=5, C=1); 14 mi SW Silver City (T=1, F=1, C=1); Tyrone (T=1, M=1).

CHIRICAHUA MOUNTAINS (58)

ARIZONA: Cochise County: Apache (T=3, M=3, C=2); Chiricahua Mts. (T=2, M=1, C=1); Cave Cr. drainage, 6200 ft, 3.5 mi S & 5.5 mi W Portal, Chiricahua Mts. (T=3, M=2, F=1, E=3); Pine Canyon, 5000 ft, Chiricahua Mts. (T=1, F=1); Paradise (T=11, M=5, F=5, C=2); between Portal and Paradise (T=1, F=1); Portal, 5000 ft, Chiricahua Mts. (T=1, F=1); 2 mi SE Portal, 4800 ft, Chiricahua Mts. (T=2, M=2); Rucker Canyon, Chiricahua Mts. (T=1, F=1); NE slope Silver Pk., 5400 ft, 0.5 mi S & 1.75 mi E Paradise, Chiricahua Mts. (T=9, M=3, F=5, E=9).
NEW MEXICO: Hildago County: Clanton Canyon, Peloncillo Mts. (T=1, M=1).

SOUTH-CENTRAL NEW MEXICO (59)

NEW MEXICO: Lincoln County: 8000 ft, Alto (T=7, M=4); Ancho (T=26, M=16, F=10, C=1); 2.5 mi S & 4.5 mi E Ancho, 6900 ft, Jicarilla Mts. (T=16, M=10, F=5, E=16); Capitan (T=1, M=1). **Sierra County:** Salinas Pk., San Andres Mts. (T=1, F=1, C=1). **Socorro County:** French Ranch, 27 mi NW Carrizozo (T=1, F=1).

GUADALUPE MOUNTAINS (60)

TEXAS: Culberson County: Frijole, 5500-6000 ft (T=3, M=1, F=2, C=2); Frijole, Nipple Hill, 5400 ft [Guadalupe Mts.] (T=2, M=2, C=1); Frijole, Pine Springs Canyon, 5800-5900 ft (T=2, M=1, F=1, C=2); Guadalupe Mts. Natl Park, Frijole (T=1, F=1).
NEW MEXICO: Eddy County: Guadalupe Mts. (T=1); Queen, 6500 ft (T=1, F=1).

MAPPED LOCALITIES EXCLUDED FROM SAMPLE AREAS

Localities of specimens plotted in Figs. 3-4 but outside of the delineated sample areas are listed below. The total number of specimens from each locality is indicated by the letter "T."
ARIZONA: Apache County: Springerville (T=1). **Coconino County:** Supai (T=1). **Gila County:** Lower Parker Cr., Sierra Ancha mts. (T=1). **Graham County:** Stanley (T=1). **Greenlee County:** 26 mi N Clifton [State Hwy. 81, Coronado Trail] (T=1); Maley Gap, Eagle Cr. drainage, San Carlos Indian Reservation (T=1); Mitchell Pk., 7600 ft (T=2). **Mojave County:** 2.5 mi ENE Aspen Pk., Hualapai Mts. (T=1). **Pima County:** near Tucson, below Bear Canyon, Santa Catalina Mts. (T=1). **Yavapai County:** Burro Cr., 3500 ft, 2 mi E Mojave Co. boundary (T=1); Tonto Basin (T=1); above Walnut Cr. work station, hill between Walnut & Apache creeks (T=1).
CALIFORNIA: Colusa County: Colusa (T=1). **Imperial County:** Vail Ranch, 5.5 mi N Westmoreland (T=1). **Inyo County:** Fall Canyon, 6500 ft, Grapevine Mts. (T=2); Johnson Canyon, ca. 6000 ft, Panamint Mts. (T=2); Panamint Mts., 7500-8000 ft (T=2); Walker Cr.,

4 mi SW Olancha [5200 ft] (T=4); 3 mi W & 4 mi S Olancha, 5700 ft (T=1). **Kern County:** Bakersfield (T=1); 14.5 mi W, 6.5 mi N McKittrick, 3500 ft (T=3); 8 mi W, 3 mi N McKittrick, 2100 ft (T=4); Tehachapi (T=1). **Merced County:** Pacheco Pass [summit] (T=2). **Modoc County: 1.25 mi S & 2 mi E Caldwell Butte, 4300 ft** (T=3); **1.75 mi S & 2 mi E Caldwell Butte, 4300 ft** (T=1); **0.75 mi N & 1.5 mi E East Sand Butte, 4100 ft** (T=2). **Sacramento County:** 0.5 mi S Sacramento State Univ., Sacramento city, W side American R. (T=1). **Santa Barbara County:** 20 mi ENE Santa Maria (T=1). **Santa Clara County:** Isabel Valley (T=1); Mt. Hamilton (T=1); 3 mi N Gerber Ranch, 2400 ft, San Antonio Valley (T=1); Smith's Cr. (T=1). **Shasta County: 3000 ft, 5.5 mi S & 4.5 mi W Fall River Mills** (T=9); **3200 ft, 5.5 mi S & 0.5 mi W Fall River Mills** (T=5); Mt. Lassen (T=1). **Siskiyou County: 1.5 mi S & 1.5 mi E Caldwell Butte, 4400 ft, near SE entrance Lava Beds Natl Monument** (T=3); **1.5 mi S & 2 mi E Caldwell Butte, 4400 ft, near SE entrance Lava Beds National Monument** (T=1); Scott R., 6 mi NW Callahan (T=1). **Trinity County:** 1 mi W Hyampom, 1200 ft (T=2). **Tulare County:** Porterville (T=11); Tule R., 8 mi E Porterville (T=1). **Yolo County:** Grafton (T=2); Woodland (T=2).

COLORADO: Mesa County: 9 mi S DeBeque (T=1). **Moffat County:** Escalante Hills, 20 mi SE Ladore (T=1). **Montrose County:** Coventry, 6800 ft (T=1).

NEVADA: Churchill County: 6800 ft, 1.75 mi S Fairview Pk. (T=1). **Clark County:** Cedar Basin, Virgin Mts., 3500 ft (T=1); 0.75 mi E & 0.75 mi S Virgin Pk., 6100 ft, Virgin Mts. (T=1). **Elko County:** 9 mi NE Wells, 6000 ft (T=2). **Esmeralda County:** 1 mi SW Indian Spring, 7500 ft, NW slope Mt. Magruder (T=2); 0.5 mi S Pigeon Springs, 6400 ft (T=1). **Eureka County: S Fork Allison Cr., 7100 ft, E slope Monitor Range** (T=1). **Mineral County: Dunlap Canyon, 7000 ft, 2.25 mi N Pilot Pk., Pilot Mts.** (T=1). **Nye County:** 7500 ft, 4 mi N Silverbow, Kawich Mts. (T=1). **Pershing County: Raspberry Cr. Canyon, 5400 ft, 1 mi S & 2.25 mi W Dun Glen Pk., East Range** (T=4). **Washoe County:** E side Granite Mtn. (T=2). **White Pine County:** Ely (T=1); Lehman Cr., 7500 ft (T=1); Sacramento Pass, 7154 ft, Snake Range (T=2); Snake Cr., 6600 ft, Snake Range (T=1).

NEW MEXICO: Catron County: Pleasanton (T=1). **Dona Ana County:** Ash Springs, San Andres Nat'l Wildlife Refuge (T=1); Soledad Canyon, 6500 ft, Organ Mts. (T=2). **Grant County:** Little Dry Cr. between Cliff & Glenwood (T=1). **Socorro County:** S slope Gallinas Pk., 7500 ft (T=2); 10 mi W & 3.5 mi S Magdalena (T=1); Riley (T=1).

OREGON: Harney County: Steens Mts., Blitzen Canyon (T=1).

UTAH: Beaver County: Pine Grove (Pine Canyon), 7200 ft, Wah Wah Mts. (T=1); Pine Canyon, 6200 ft, Wah Wah Mts., 25 mi W Milford (T=1); Pine Canyon, 7 mi E from turnoff, 6800 ft, Wah Wah Mts. (T=1). **Box Elder County:** 3 mi NW Promontory Station (T=1). **Carbon County:** Sunnyside (T=1). **Daggett County:** Hideout Forest Camp on Green R., 5900 ft (T=2). **Grand County:** 4 mi S Flatrock Jcn., E Tavaputs Plateau, 7600 ft [40 mi S Ouray] (T=3); Willow Springs, near Arches Nat'l Monument (T=1). **Juab County:** S end Deep Cr. Mts., 5500 ft, 11 mi W Trout Cr., near Utah-Nevada border (T=1). **Kane County:** Kaiparowitz Plateau, 7000 ft (T=3). **San Juan County:** Natural Bridges (T=1). **Uintah County:** Cottonwood Springs, 8 mi W Vernal (T=3); 1 mi W Flatrock Jcn., 40 mi S Ouray, E Tavaputs Plateau, 7500 ft (T=3); 2 mi N Pine Spring, 38 mi SE Ouray, 7000 ft (T=1); W margin Wolf Flat, E Tavaputs Plateau, 7200 ft (T=1). **Utah County:** Provo (T=1).

WYOMING: Sweetwater County: Green R., 4 mi NE Linwood, Utah (T=1).

MEXICO: Sonora: W slope San Luis Mts. (T=4).

OTHER LOCALITIES WITH SPECIMENS

Specimens of *P. inornatus* were examined from several localities that either (a) could not be found on any maps or in any gazetteers, (b) were too general to plot, or, (c) in one case, represent an extremely unusual record. The total number of specimens from each locality is indicated by the letter "T."

ARIZONA: Navajo County: mouth of Waterlily Canyon, 6500 ft (T=1).

CALIFORNIA: Alameda County: no locality (T=4). **Butte County:** no locality (T=1). **Glenn County:** Stonyford Cr. (T=1). **Los Angeles County:** no locality (T=3). **Riverside County:** Coleman Tin Mine [1275 ft] [Cleveland Nat'l Forest] (T=2). **San Bernardino County:** 18 mi SE Victorville (T=1); no locality (T=3). **San Diego County:** Moro Pk. (T=2); no locality (T=1). **Santa Clara County:** no locality (T=1). **Santa Cruz County:** no locality (T=2). **Sonoma County:** no locality (T=2). **Ventura County:** no locality (T=1). **County Unknown:** Mt. George (T=1); Sacramento Valley (T=3); no locality (T=2).

COLORADO: Baca County: Jimmie Cr. (T=1). **Gunnison County:** Dayton (T=1). **La Plata County:** Animas R. (T=2). **Moffat County:** no locality (T=1). **Routt County:** Douglas Spring (T=1). **County Unknown:** Eurango [= Durango?] (T=1); Rio Almos (T=1); no locality (T=2).

NEVADA: Washoe County: Pyramid Lake (T=1); no locality (T=2). **County unknown:** along Hwy. 50 near Churchill-Ormsby [= Carson City] county line (T=1) [these two counties do not border each other].

NEW MEXICO: Socorro County: Cooper's Well, 6900 ft, 33°56'N, 106°12'W (T=2). **County Unknown:** S.O.S. Ranch (T=1); no locality (T=1).

UTAH: Beaver County: Beaver Canyon (T=1). **Millard County:** Paxton's Ranch (T=1). **San Juan County:** Moab to Monticello, 3900 ft (T=1); Navajo Mtn. Trading Post, 6500 ft (T=1).

WASHINGTON: Yakima County: Yakima (T=1). This specimen, a female, was collected by G. G. Cantwell on 10 August 1911 (specimen number 11765, University of California, Los Angeles). Because the locality is at least 300 mi from the nearest known breeding population in southern Oregon, and given the sedentary behavior of *P. inornatus*, this record is suspect. While Cantwell supposedly was a reliable collector (J. Northern, *in litt.*, 18 July 1989; K. Garrett, *in litt.*, 24 October 1989), field notes or other evidence are needed for confirmation.

LOCALITIES OF OUTGROUP SPECIMENS

Tissue samples for outgroup analysis were obtained from the frozen collection of the Museum of Vertebrate Zoology, University of California, Berkeley. Localities of outgroup specimens are listed below.

Parus bicolor: **Oklahoma: Haskell County:** 6 mi N & 1 mi W McCurtain, 480 ft (T=2, E=2); 4 mi NE Stigler, 650 ft (T=1, E=1).

Parus gambeli: **California: Trinity County:** 0.5 mi S & 1 mi W Scott Mtn. summit, 1620 m (T=5, E=5).

Appendix B

Revised Taxonomy and Nomenclature

This appendix discusses the revised taxonomy of *Parus inornatus* and *P. ridgwayi* based on recommendations resulting from the current study (see text and Fig. 45). Nomenclatural synonymies are provided for each taxon, along with details on type specimens and geographic distributions. Information is organized by subspecies, and the nomenclatural history of these forms is reviewed. Previous notions concerning variation in the *Parus inornatus* complex are discussed and interpreted in light of patterns detected in this study through the use of modern systematic techniques.

Parus inornatus inornatus Gambel

Parus inornatus Gambel (1845:265), original description.
Baeolophus inornatus restrictus Ridgway (1903:109).
Baeolophus inornatus inornatus Ridgway (1904:387-389).
Baeolophus inornatus sequestratus Grinnell and Swarth (1926:166-168).
Baeolophus inornatus kernensis Grinnell and Behle (1937:225-226).

Type. Specimen number 3340, United States National Museum; sex not indicated; subadult; "Upper California" (= Monterey, Monterey County, California; *fide* Gambel 1847:154); probably 20 November 1842; collected by W. Gambel.
Distribution. Southwestern Oregon (Upper Rogue River Valley, Jackson County) and northwestern and central California to Santa Barbara and Kern counties. Ranges southward along the Coast, Transverse, and Peninsular ranges, and in the western foothills of the Sierra Nevada. Intergrades with *Parus inornatus affabilis* in Santa Barbara and Ventura counties and at the southern end of the Sierra Nevada.

177

Remarks. The American Ornithologists' Union (1957) did not recognize *Baeolophus inornatus restrictus*, which was named by Ridgway (1903:109) for titmice restricted to the San Francisco Bay Area (type: specimen no 163569, United States National Museum; male; adult; Oakland, Alameda County, California; 24 March 1896; collected by J. Hornung). This race was described as "similar to *B. i. inornatus* but darker, especially the underparts, the young conspicuously less brownish" (ibid.).

Grinnell and Swarth (1926:166-168) described *Baeolophus inornatus sequestratus* from specimens collected in Jackson County, Oregon (type: specimen number 46163, Museum of Vertebrate Zoology; male; Eagle Point, Jackson County, Oregon; 26 November 1925; collected by William E. Sherwood). The distribution of this form included populations from southwestern Oregon and northwestern California, including the valley of the South Fork of the Trinity River and "probably.... scatteringly in intervening valleys" (Grinnell and Miller 1944:307). Isolated from *P. i. inornatus* to the south by coniferous forest near Mt. Shasta, populations in northwestern California also were reported to be separated from "*P. i. griseus*" (= *P. ridgwayi zaleptus*) to the east by "coniferous forest or sagebrush plain" (Grinnell and Swarth 1926:166). However, recent field work on the Modoc Plateau in northern California indicates that populations of *inornatus* in this region extend eastward at least to Mt. Dome, where they exhibit tenuous secondary contact with interior *zaleptus* (see text for detailed discussion). Although some Pacific slope populations exhibit interior traits in habitat preference and allelic frequencies, the overall similarity of *sequestratus* to *inornatus* warrants merging it into the nominate form.

In 1937, Grinnell and Behle (1937:225-226) recognized another differentiate of *P. i. inornatus* (*P. i. kernensis*) in the Kern River drainage basin on the southwest slope of the Sierra Nevada (type: specimen number 63801, Museum of Vertebrate Zoology; male; adult; Rankin Ranch, 3300 ft, Walker Basin, Kern County, California; 19 November 1933; collected by R. M. Gilmore, orig. no. 3235). These authors distinguished *P. i. kernensis* from *P. i. inornatus* by its more grayish plumage, noting that the color is intermediate between coastal and Great Basin titmice. According to Grinnell and Behle (ibid.), however, "it is doubtful whether direct intergradation with any of the latter [i.e., interior] races actually exists because of the great disparity in bill size, unabridged by any specimen seen." My analyses of both phenotypic and genetic variation strongly support this conclusion. On the other hand, the data indicate that *kernensis* is not distinct from *inornatus* and does not warrant separate subspecies status.

Parus inornatus affabilis (Grinnell and Swarth)

Baeolophus inornatus griseus Ridgway (Bryant 1889:317), part.
Baeolophus inornatus murinus Ridgway (1903:109), original description.
Baelophus inornatus affabilis Grinnell and Swarth (1926:164-166), part.
Baeolophus inornatus transpositus Grinnell (1928:154), part.
Baeolophus inornatus murinus Ridgway (Grinnell 1928:154), part.

Type. Specimen number 47074, Museum of Vertebrate Zoology; male; Concepcion, 6000 ft, Sierra San Pedro Mártir, Lower California, Mexico; 20 November 1925; collected by C. C. Lamb, orig. no. 5278.

Distribution. Southwestern California and northern Baja California, primarily west of the desert divides (including the San Bernardino and San Jacinto mountains). Occurs from Santa Barbara, Ventura, and southern Kern counties, California, south to the west slope of the Sierra San Pedro Mártir, Baja California Norte. Intergrades to the north with *Parus inornatus inornatus*.

Remarks. Plain Titmice were initially found in the Sierra San Pedro Mártir by A. W. Anthony and ascribed to the form *Baeolophus inornatus griseus* (Bryant 1889:317; Anthony 1893:246). Ridgway (1903:109) grouped titmice from this locality with populations from Nachoguero Valley and gave the new name *murinus* to this subspecies (type: specimen no. 133812, United States National Museum; male; adult; Nachoguero Valley, Lower California; 4 June 1894; collected by E. A. Mearns). The distribution of *B. i. murinus* was described as "Southern California, in Los Angeles, San Bernardino, and San Diego counties; northern Lower California" (ibid.). Grinnell and Swarth (1926:164-166) later studied specimens from the Sierra San Pedro Mártir and named a new subspecies of titmouse (*B. i. affabilis*) with its center of differentiation in that region. Because Nachoguero Valley lies just south of the U.S.-Mexico boundary (between Campo and Jacumba), these workers presumed that birds from that locality would be closest to those from the San Diegan subfaunal district. Thus, the name *affabilis* was limited to titmice occurring south of Nachoguero Valley in Baja California Norte (e.g., in the Sierra San Pedro Mártir and Sierra Juarez), and *murinus* was applied to birds from southern California. Subsequently, Chester C. Lamb was commissioned by the Museum of Vertebrate Zoology to collect topotype series of several species from Nachoguero Valley. Included in this collection was a series of titmice in fresh fall plumage. Upon examination of this material, Grinnell (1928:154) concluded that the birds from Nachoguero Valley were more similar to titmice from the Sierra San Pedro Mártir than to those from the "San Diegan subfaunal district" despite the geographic proximity of the latter: "While this series is not quite extreme of the San Pedro Mártir race, it is in characters on that side of the median line in the belt of intergradation between the San Diegan race and the San Pedro Mártir one." Accordingly, Grinnell (ibid.) proposed the name *murinus* for the subspecies from northern Baja California, and gave the new name *transpositus* to titmice from southern California (type: specimen no. 38685, Museum of Vertebrate Zoology; male; Mount Wilson, Los Angeles County, California; December 12, 1896; collected by J. Grinnell, orig. no. 1828). The subspecific name *affabilis* was re-established because *murinus* was preocuppied by *Parus murinus* Brehm, 1855 (fide Paynter 1967:121). After examining specimens from numerous localities in southern California and northern Baja California, I found no basis for recognizing two subspecies in this region. Thus, I propose that all titmice currently ascribed to the subspecies *P. i. transpositus* and *P. i. affabilis* be combined under the single form *affabilis*.

Parus inornatus mohavensis Miller

Parus inornatus mohavensis A.H. Miller (1946:76-78), original description.

Type. Specimen number 94208, Museum of Vertebrate Zoology; male; adult; Pinyon Wells, 4000 ft, Little San Bernardino Mountains, Riverside County, California; 12 October 1945; collected by A. H. Miller, orig. no. 5581.

Distribution. Restricted to the Upper Sonoran Zone of the Little San Bernardino Mountains, San Bernardino and Riverside counties, California (Joshua Tree National Monument). Occurs from Morongo Valley eastward to the vicinity of Little San Bernardino Mountain (north of Mecca). Not found on Eagle Mountain, an isolated peak southeast of the Little San Bernardino Mountains (A.H. Miller 1946:76; Peterson 1990:127-135).

Remarks. A.H. Miller (1946:76-78) described this subspecies primarily on the basis of dorsal coloration, which he remarked as being grayer than other coastal races with the exception of *cineraceus* in southern Baja California. Miller presumed that the loss of brown reflects adaptation of a coastal form to the arid environment of the region. However, he also suggested that this modification, in conjunction with an increase in average bill size, may have resulted from occasional intergradation with gray titmice of the interior (e.g., "*Parus inornatus ridgwayi*" from the Providence Mountains in eastern San Bernardino County). Nonetheless, Miller acknowledged that the large gap (approx. 50 mi) of unsuitable desert habitat between *mohavensis* and *ridgwayi* would be "a formidable barrier for a strictly resident species" (ibid.:76). Furthermore, he argued that "*mohavensis* shows no exceptionally great amplitude of individual variation [i.e., in bill size] such as often occurs in areas of secondary intergradation" (ibid.:77). According to Miller (ibid.:76), *mohavensis* probably attains some continuity with "*transpositus*" (= *affabilis*) across Morongo Pass on the east slope of the massive San Bernardino Mountains.

The genetic results of my study support Miller's contention that *P. i. mohavensis* is derived from Pacific slope populations to the west. Although the phenetic break between *mohavensis* and *P. i. affabilis* is not as strong as between other subspecies, the differences in color are sufficient to warrant continued recognition of *mohavensis* as a separate form of *Parus inornatus*.

Parus inornatus cineraceus (Ridgway)

Lophophanes inornatus cineraceus Ridgway (1883:154-155), original description.
Baeolophus inornatus cineraceus Ridgway (1904:391-392).

Type. Specimen number 89800, United States National Museum; male; adult; Laguna (= La Laguna), Lower California; 2 February 1883; collected by L. Belding.

Distribution. Restricted to the mountains in the Cape District, Baja California Sur (Victoria Mountains, Sierra de la Laguna).

Parus ridgwayi zaleptus (Oberholser)

Lophophanes inornatus griseus Ridgway (1882:344), part.
Parus inornatus ridgwayi Richmond (1902:155), part.
Baelophus inornatus griseus Ridgway (1904:390), part.
Baeolophus inornatus zaleptus Oberholser (1932:7-8), original description, part.

Type. Specimen number 15008, Cleveland Museum of Natural History; female; adult; rim of Warner Valley northwest of the Jacobs Ranch, Twenty Mile Creek, 9 mi S Adel, [Lake County], Oregon; 3 May 1930; collected by A. Walker, orig. no. 2231.

Distribution. Southeastern Oregon (Warner Valley and Hart Mtn., Lake County), northeastern and eastern California (south to Providence Mountains), and western to eastern and southeastern Nevada. Rare or absent in most mountain ranges of central Nevada. Exhibits tenuous secondary contact with *Parus inornatus inornatus* on the Modoc Plateau in northern California.

Remarks. Titmice from the interior regions of California were originally lumped with other Great Basin and southwestern populations under the name *Baelophus inornatus griseus* (= "*Parus inornatus ridgwayi*"), the "Gray Titmouse." Grinnell (1923:135-137), in a review of the distribution and existing records of "*griseus*" in California, commented on the striking differences between this form and coastal subspecies recognized at the time (*inornatus* and "*murinus*" [= *affabilis*]). A comparison of specimens of "*griseus*" from eastern California with those from northern Arizona revealed no differences. However, the following comments (ibid.:136-137) underscore the strong dissimilarity noted by Grinnell between "*griseus*" and the "*inornatus-murinus*" complex:

> With regard to distinctness, my study of the series of skins available at this writing.... leaves me with the strong impression that the Gray Titmouse is set off much more sharply from the *inornatus-murinus* titmouses than has hitherto been supposed. In spite of statements and implications to the effect that intergradation between *inornatus* and *griseus* occurs in the region of the southern Sierra Nevada, I have failed to find even one intermediate.... Furthermore,.... information so far available indicates a geographic hiatus between the range of *griseus* and the range of *inornatus*.... I think it very unlikely that there is any well-marked continuity of favorable conditions there [on the east flank of the Sierra Nevada north from Kern County to Mono County], such as would have to be present to permit free intergradation.... *griseus* belongs to the pinyon-juniper association; *inornatus* to the analogous digger-pine [gray pine] and oak association.... I am almost tempted to propose full specific status for the Gray Titmouse.

Concerning the area of possible intergradation between "*griseus*" and *inornatus*, Grinnell (ibid.:137) remarked on the paler color of titmice in the southern Sierra Nevada relative to other coastal populations. Although this difference in color was used later to characterize the subspecies "*kernensis*" (Grinnell and Behle 1937:225-226), Grinnell (1923:137) noted that the "paleness consists merely in lightening of the tone of brown dorsally and a whitening of the lower surface; it does not tend toward the leaden hue both above and below characteristic of "*griseus*."

Grinnell and Swarth (1926:168) also emphasized the probable independence of "*griseus*":

> It approaches the range of the Oregon Plain Titmouse [= "*sequestratus*"] fairly closely, and there are places where it still nearer approaches the habitat of *inornatus*; but in each case the nature of the intervening territory is such as to render it highly improbable that there is actual meeting, and consequent intergradation, of these subspecies at any point.

The recent finding that "*sequestratus*" inhabits pure juniper in addition to oaks or oaks mixed with juniper refutes the hypothesis that these forms are ecologically isolated.

The subspecies *zaleptus* was originally described by Oberholser (1932:7-8) as "Similar to *Baeolophus inornatus griseus*, but much more clearly grayish above with practically none of the brownish tinge so evident in the latter race...." Linsdale (1938a:37-38) compared freshly plumaged specimens of *zaleptus* from northeastern California and northwestern Nevada with specimens of *ridgwayi* from eastern Nevada, and concluded that *zaleptus*, although distinctive, differs in being darker rather than paler in color. Linsdale (ibid.:38) also remarked on the large bill of *zaleptus* ("larger and broader in these birds than in *griseus* or any other form of the species") and stated that "This race cannot be considered intermediate for, in several characters, it reaches extreme development for the species." Although *zaleptus* was not known to intergrade with *ridgwayi*, Grinnell and Miller (1944:306) noted that "it is possible that populations toward the south, in Mono and Inyo counties, are more properly referable to the race *P. i. ridgwayi* than to *zaleptus*."

Whereas I found no clear phenotypic patterns differentiating *zaleptus* and *ridgwayi* as discrete entities, the analysis of allozymes clearly linked populations of titmice from eastern and southeastern Nevada (e.g., at least to the Ruby Mountains and Cedar Range) with those from eastern California and southeastern Oregon. Consequently, I propose that all of these populations be combined under the subspecies name *zaleptus*. Although individuals from the Eastern Mojave (e.g., the Providence Mountains) were more similar electrophoretically to those from the Pacific slope, they are interior in terms of phenotype, habitat preference, and distribution and thus are ascribed to the form *zaleptus*.

Parus ridgwayi ridgwayi Richmond

Lophophanes inornatus griseus Ridgway (1882:344), original description.
Parus inornatus griseus Ridgway (1882:344), part.

Parus inornatus ridgwayi Richmond (1902:155), part.
Baeolophus inornatus griseus Ridgway (1904:390-391),part.
Baeolophus inornatus plumbescens Grinnell (1934:251-252).

Types. This name is based on three cotypes in the United States National Museum which are the only ones that "belong to *griseus* in the strictest sense" (Deignan 1961:346-347): Specimen number 62856; male; adult; "Iron City," Iron County, Utah; 8 October 1872; collected by H. W. Henshaw, orig. no. 263. Specimen number 68791; male; adult; Santa Fe, Santa Fe County, New Mexico; 16 January 1874; collected by H. W. Henshaw, orig. no. 41A. Specimen number 69403; sex not indicated; adult; El Paso County, Colorado; 14 January 1874; collected by C. E. H. Aiken, orig. no. 377. The type locality from Utah is the one currently recognized by the American Ornithologists' Union (1957:393).

Distribution. Southeastern Idaho and northwestern Utah to southwestern Wyoming, south-central Colorado, southeastern Arizona, south-central New Mexico, and western Oklahoma. Barely extends into northwestern Chihuahua, Mexico (San Luis Mountains). Also occurs in the Guadalupe Mountains of southern New Mexico and northwestern Texas.

Remarks. Ridgway (1882:344) based the form *Lophophanes inornatus griseus* on an extensive series of specimens from Colorado, Utah, Nevada, Arizona, and New Mexico. Subsequently, some of these specimens were assigned by other workers to the subspecies *zaleptus* or *plumbescens*. Ridgway's taxon was renamed by Richmond (1902:155) as *Parus inornatus ridgwayi* because the name *griseus* was preoccupied by *Parus griseus* Gmelin, 1789.

The name *plumbescens* (Grinnell 1934:251-252) was described from southwestern specimens of *ridgwayi* (type: Specimen number 65010, Museum of Vertebrate Zoology; male; Silver City, Grant County, New Mexico; 29 March 1933; collected by A. Brooks, orig. no. 7373). This subspecies was thought to occur from central and southeastern Arizona (south of the Colorado and Little Colorado rivers) to southwestern New Mexico and northwestern Chihuahua, Mexico. Phillips et al. (1964) did not recognize *plumbescens* in Arizona, and ascribed the name *ridgwayi* to all interior populations of titmice. They based this argument on the view that "The purely gray, interior Plain Titmice show a mosaic of non-clinal geographic variations in size and darkness of the gray tones (Phillips, Jour. Ariz. Acad. Sci. 1, 1959: 28).... Therefore no true races can be defined...." (ibid.:111). Although I was unable to locate this original reference, I agree with their conclusion that there is no clear pattern of variation differentiating *ridgwayi* and *plumbescens*. While size tends to increase toward the north, I found no evidence for a phenotypic or genetic break between these forms. Therefore, recognition of *plumbescens* and *ridgwayi* as separate units is not justified. As discussed previously, *zaleptus* is recognizable based on allozymes, even though size and color data fail to separate it from either "*plumbescens*" or *ridgwayi*.

Literature Cited

Alatalo, R. V., and L. Gustafson
   1988   Genetic component of morphological differentiation in Coal Tits under competitive release. Evol. 42:200-203.

Alcorn, J. R.
   1988   The Birds of Nevada. Fairview West Publishing, Fallon, Nev.

Aldrich, J. W.
   1984   Ecogeographical variation in size and proportions of Song Sparrows (*Melospiza melodia*). Ornithol. Monogr. 35:1-134.

Aldrich, J. W., and F. C. James
   1991   Ecogeographic variation in the American Robin (*Turdus migratorius*). Auk 108:230-249.

Amadon, D.
   1959   The significance of sexual differences in size among birds. Proc. Amer. Phil. Soc. 103:531-536.

Amadon, D., and L. L. Short
   1992   Taxonomy of lower categories—suggested guidelines. Bull. Brit. Ornithol. Club 112A:11-38.

American Ornithologists' Union
   1957   Check-list of North American Birds, 5th ed. American Ornithologists' Union, Washington, D.C.

1983   Check-list of North American Birds, 6th ed. American Ornithologists' Union, Washington, D.C.

1988   Report of American Ornithologists' Union, Cooper Ornithological Society, and Wilson Ornithological Society *ad hoc* committee on the use of wild birds in research. Auk 105 (suppl.):1a-41a.

Anthony, A. W.
1893   Birds of San Pedro Mártir, Lower California. Zoe 4:228-247.

Antonovics, J.
1971   The effects of a heterogeneous environment on the genetics of natural populations. Amer. Sci. 59:593-599.

Arnold, M. L.
1992   Natural hybridization as an evolutionary process. Ann. Rev. Ecol. Syst. 23:237-261.

Atwood, J. L.
1988   Speciation and geographic variation in Black-tailed Gnatcatchers. Ornithol. Monogr. 42:1-74.

Avise, J. C.
1994   Molecular Markers, Natural History and Evolution. Chapman and Hall, New York.

Avise, J. C., J. Arnold, R. M. Ball, E. Bermingham, T. Lamb, J. E. Neigel, C. A. Reeb, and N. C. Saunders
1987   Intraspecific phylogeography: The mitochondrial DNA bridge between population genetics and systematics. Ann. Rev. Ecol. Syst. 18:489-522.

Avise, J. C., and R. M. Ball, Jr.
1990   Principles of genealogical concordance in species concepts and biological taxonomy. Oxford Surv. Evol. Biol. 7:45-67.

Avise, J. C., and R. M. Zink
1988   Molecular genetic divergence between avian sibling species: King and Clapper Rails, Long-billed and Short-billed Dowitchers, Boat-tailed and Great-tailed Grackles, and Tufted and Black-crested Titmice. Auk 105:516-528.

Axelrod, D. I.

1948 Climate and evolution in western North America during middle Pliocene time. Evol. 2:127-144.

1958 Evolution of the Madro-Tertiary geoflora. Bot. Rev. 24:433-509.

1973 History of the Mediterranean ecosystem in California. In F. di Castri and H. A. Mooney (eds.), Mediterranean Type Ecosystem: Origin and Structure, pp. 225-277. Springer-Verlag, Berlin.

1977 Outline history of California vegetation. In M. G. Barbour and J. Major (eds.), Terrestrial Vegetation of California, pp. 139-193. John Wiley, New York.

1979 Age and origin of Sonoran Desert vegetation. Occas. Papers Calif. Acad. Sci. 132:1-74.

1981 Holocene climatic changes in relation to vegetation disjunction and speciation. Amer. Nat. 117:847-870.

1983 Paleobotanical history of the western deserts. In S. G. Wells and D. R. Haragan (eds.), Origin and Evolution of Deserts, pp. 113-129. University of New Mexico Press, Albuquerque.

Baker, A. J.

1992 Genetic and morphometric divergence in ancestral European and descendant New Zealand populations of chaffinches (*Fringilla coelebs*). Evol. 46:1784-1800.

Baker, A. J., M. K. Peck, and M. A. Goldsmith

1990 Genetic and morphometric differentiation in introduced populations of Common Chaffinches (*Fringilla coelebs*) in New Zealand. Condor 92:76-88.

Baker, M. C., and A. E. M. Baker

1990 Reproductive behavior of female buntings: Isolating mechanisms in a hybridizing pair of species. Evol. 44:332-338.

Baldwin, S. P., H. C. Oberholser, and L. G. Worley

1931 Measurements of birds. Cleveland Mus. Nat. Hist. Sci. Publ. 2:1-165.

Ball, R. M., Jr., and J. C. Avise

1992 Mitochondrial DNA phylogeographic differentiation among avian populations and the evolutionary significance of subspecies. Auk 109:626-636.

Ball, R. M., Jr., S. Freeman, F. C. James, E. Bermingham, and J. C. Avise
    1988   Phylogeographic population structure of Red-winged Blackbirds assessed by mitochondrial DNA. Proc. Natl. Acad. Sci. USA 85:1558-1562.

Banks, R. C.
    1964   Geographic variation in the White-crowned Sparrow, *Zonotrichia leucophrys.* Univ. Calif. Publ. Zool. 70:1-123.
    1986   Subspecies of the Glaucous Gull, *Larus hyperboreus* (Aves: Charadriiformes). Proc. Biol. Soc. Wash. 99:149-159.
    1988   Geographic variation in the Yellow-billed Cuckoo. Condor 90:473-477.

Barrowclough, G. F.
    1980a   Gene flow, effective population sizes, and genetic variance components in birds. Evol. 34:789-798.
    1980b   Genetic and phenotypic differentiation in a wood warbler (genus *Dendroica*) hybrid zone. Auk 97:655-668.
    1982   Geographic variation, predictiveness, and subspecies. Auk 99:601-603.
    1983   Biochemical studies of microevolutionary processes. In A. H. Brush and G. A. Clark (eds.), Perspectives in Ornithology, pp. 223-261. Cambridge University Press, Cambridge, Eng.
    1992   Book review: Speciation and geographic variation in Black-tailed Gnatcatchers. Condor 94:555-556.

Barrowclough, G. F., and R. J. Gutiérrez
    1990   Genetic variation and differentiation in the Spotted Owl (*Strix occidentalis*). Auk 107:737-744.

Barrowclough, G. F., and N. K. Johnson
    1988   Genetic structure of North American birds. Acta Congr. Intern. Ornithol. 19:1630-1638, 1669-1673.

Barrowclough, G. F., N. K. Johnson, and R. M. Zink
    1985   On the nature of genic variation in birds. Current Ornithol. 2:135-154.

Barton, N. H., and B. Charlesworth
    1984   Genetic revolutions, founder effects, and speciation. Ann. Rev. Ecol. Syst. 15:133-164.

Barton, N. H., and G. M. Hewitt
    1985   Analysis of hybrid zones. Ann. Rev. Ecol. Syst. 16:113-148.

Bates, J. M., and R. M. Zink
   1992   Seasonal variation in gene frequencies in the House Sparrow (*Passer domesticus*). Auk 109:658-662.

Behle, W. H.
   1942   Distribution and variation of the Horned Larks (*Octocoris alpestris*) of western North America. Univ. Calif. Publ. Zool. 46:205-316.
   1956   A systematic review of the Mountain Chickadee. Condor 58:51-70.
   1963   Avifaunistic analysis of the Great Basin region of North America. Proc. Intern. Ornithol. Congr. 13:1168-1181.

Bell, D. A.
   1992   Hybridization and sympatry in the Western Gull/Glaucous-winged Gull complex. Ph.D dissertation, University of California, Berkeley.

Bent, A. C.
   1946   Life histories of North American jays, crows, and titmice. U.S. Natl. Mus. Bull. 191:1-495.

Bermingham, E., S. Rohwer, S. Freeman, and C. Wood
   1992   Vicariance biogeography in the Pleistocene and speciation in North American wood warblers: A test of Mengel's model. Proc. Natl. Acad. Sci. USA 89:6624-6628.

Betancourt, J. L., T. R. Van Devender, and P. S. Martin
   1990   Packrat Middens: The Last 40,000 Years of Biotic Change. University of Arizona Press, Tucson.

Bibb, M. J., R. A. Van Etten, C. T. Wright, M. W. Walberg, and D. A. Clayton
   1981   Sequence and gene organization of mouse mitochondrial DNA. Cell 26:167-180.

Bird, J., B. Riska, and R. R. Sokal
   1981   Geographic variation in variability of *Pemphigus populicaulis*. Syst. Zool. 30:58-70.

Birt-Friesen, V. L., W. A. Montevecchi, A. J. Gaston, and W. S. Davidson
   1992   Genetic structure of Thick-billed Murre (*Uria lomvia*) populations examined using direct sequence analysis of amplified DNA. Evol. 46:267-272.

Bledsoe, A. H.

    1987   Estimation of phylogeny from molecular distance data: The issue of variable rates. Auk 104:563-565.

Block, W. M.

    1989   Spatial and temporal patterns of resource use by birds in California oak woodlands. Ph.D dissertation, University of California, Berkeley.

    1990   Geographic variation in foraging ecologies of breeding and nonbreeding birds in oak woodlands. Stud. Avian Biol. 13:264-269.

Bowers, D. E.

    1956   A study of methods of color determination. Syst. Zool. 5:147-160, 182.

    1960   Correlation of variation in the Wrentit with environmental gradients. Condor 62:91-120.

Braun, M. J., and M. B. Robbins

    1986   Extensive protein similarity of the hybridizing chickadees *Parus atricapillus* and *P. carolinensis*. Auk 103:667-675.

Britten, R. J.

    1986   Rates of DNA sequence evolution differ between taxonomic groups. Science 231:1393-1398.

Brown, W. M., M. George, Jr., and A. C. Wilson

    1979   Rapid evolution of animal mitochondrial DNA. Proc. Natl. Acad. Sci. USA 76:1967-1971.

Browning, M. R.

    1975   The distribution and occurrence of the birds of Jackson County, Oregon, and surrounding areas. U.S. Fish and Wildl. Serv., North Amer. Fauna 70:1-69.

Bryant, W. E.

    1889   A catalogue of the birds of Lower California, Mexico. Proc. Calif. Acad. Sci. 2:237-320.

Burns, K. J., and R. M. Zink

    1990   Temporal and geographic homogeneity of gene frequencies in the Fox Sparrow (*Passerella iliaca*). Auk 107:421-425.

Burson, S. L., III
  1990   Population genetics and gene flow of the Common Tern. Condor 92:182-192.

Bush, G. L.
  1975   Modes of animal speciation. Ann. Rev. Ecol. Syst. 6:339-364.

Carson, H. L., and A. R. Templeton
  1984   Genetic revolutions in relation to speciation phenomena: The founding of new populations. Ann. Rev. Ecol. Syst. 15:97-131.

Cavalli-Sforza, L. L., and A. W. F. Edwards
  1967   Phylogenetic analysis: Models and estimation procedures. Evol. 21:550-570.

Cheverud, J. M.
  1988   A comparison of genetic and phenotypic correlations. Evol. 42:958-968.

Cicero, C., and N. K. Johnson
  1992   Genetic differentiation between populations of Hutton's Vireo (Aves: Vireonidae) in disjunct allopatry. Southwest. Nat. 37:344-348.
  1995   Speciation in Sapsuckers (*Sphyrapicus*): III. Mitochondrial DNA sequence divergence at the cytochrome *b* locus. Auk 112 (in press).

Cole, K.
  1982   Late Quaternary zonation of vegetation in the eastern Grand Canyon. Science 217:1142-1145.

Coplen, T. B., I. J. Winograd, J. M. Landwehr, and A. C. Riggs
  1994   500,000-year stable carbon isotopic record from Devils Hole, Nevada. Science 263:361-365.

Corbin, K. W.
  1983   Genetic structure and avian systematics. Current Ornithol. 1:211-244.

Coyne, J. A.
  1992   Genetics and speciation. Nature 355:511-515.

Cracraft, J.
  1983   Species concepts and speciation analysis. Current Ornithol. 1:159-187.

Crow, J. F., and K. Aoki
  1984   Group selection for a polygenic behavioral trait: Estimating the degree of population subdivision. Proc. Natl. Acad. Sci. USA 81:6073-6077.

Daly, J. C., and J. L. Patton
  1986   Growth, reproduction, and sexual dimorphism in *Thomomys bottae* pocket gophers. J. Mammal. 67:256-265.

Davies, J. C., R. F. Rockwell, and F. Cooke
  1988   Body-size variation and fitness components in Lesser Snow Geese (*Chen caerulescens caerulescens*). Auk 105:639-648.

Davis, J.
  1954   Seasonal changes in bill length of certain passerine birds. Condor 56:142-149.
  1957   Determination of age in the Spotted Towhee. Condor 59:195-202.
  1959   The Sierra Madrean element of the avifauna of the Cape District, Baja California. Condor 61:75-84.
  1961   Some seasonal changes in morphology of the Rufous-sided Towhee. Condor 63:313-321.

Davis, O. K., R. S. Anderson, P. L. Fall, M. K. O'Rourke, and R. S. Thompson
  1985   Palynological evidence for early Holocene aridity in the southern Sierra Nevada, California. Quat. Res. 24:322-332.

Davis, S., B. S. Davis, and J. Davis
  1973   Some factors affecting foraging behavior of Plain Titmice. Condor 75:481-482.

Deignan, H. G.
  1961   Type specimens of birds in the United States National Museum. U.S. Natl. Mus. Bull. 221:1-718.

Desjardins, P., and R. Morais
  1990   Sequence and gene organization of the chicken mitochondrial genome. J. Mol. Biol. 212:599-634.

Detling, L. E.
  1961   The chaparral formation of southwestern Oregon, with considerations of its postglacial history. Ecol. 42:348-357.

Dhondt, A. A.
  1982   Heritability of Blue Tit tarsus length from normal and cross-fostered broods. Evol. 36:418-419.

Dixon, K. L.
  1949   Behavior of the Plain Titmouse. Condor 51:110-136.
  1950   Notes on the ecological distribution of Plain and Bridled Titmice in Arizona. Condor 52:140-141.
  1954   Some ecological relations of chickadees and titmice in central California. Condor 56:113-124.
  1955   An ecological analysis of the interbreeding of crested titmice in Texas. Univ. Calif. Publ. Zool. 54:125-206.
  1956   Territoriality and survival in the Plain Titmouse. Condor 58:169-182.
  1962   Notes on the molt schedule of the Plain Titmouse. Condor 64:134-139.
  1969   Patterns of singing in a population of the Plain Titmouse. Condor 71:94-101.
  1990   Constancy of margins of the hybrid zone in titmice of the *Parus bicolor* complex in coastal Texas. Auk 107:184-188.

Dobzhansky, T.
  1937   Genetics and the Origin of Species. Columbia University Press, New York.

Douglas, M. E., and J. A. Endler
  1982   Quantitative matrix comparisons in ecological and evolutionary investigations. J. Theor. Biol. 99:777-795.

Dyck, J.
  1992   Reflectance spectra of plumage areas colored by green feather pigments. Auk 109:293-301.

Edwards, S. V., P. Arctander, and A. C. Wilson
  1991   Mitochondrial resolution of a deep branch in the genealogical tree for perching birds. Proc. R. Soc. Lond. B 243:99-107.

Edwards, S. V., and A. C. Wilson
  1990   Phylogenetically informative length polymorphism and sequence variability in mitochondrial DNA of Australian songbirds (*Pomatostomus*). Genetics 126:695-711.

Ehrlich, P. R., and P. H. Raven
  1969   Differentiation of populations. Science 165:1228-1232.

Endler, J. A.
1977   Geographic Variation, Speciation, and Clines. Princeton University Press, Princeton, N.J.

Erickson, M. M.
1938   Territory, annual cycle, and numbers in a population of wren-tits (*Chamaea fasciata*). Univ. Calif. Publ. Zool. 42:247-334.

Farner, D. S.
1949   Age groups and longevity in the American Robin: Comments, further discussion, and certain revisions. Wilson Bull. 61:68-81.

Farris, J. S.
1972   Estimating phylogenetic trees from distance matrices. Amer. Nat. 106:645-668.
1981   Distance data in phylogenetic analysis. In V. A. Funk and D. R. Brooks (eds.), Advances in Cladistics: Proceedings of the First Meeting of the Willi Hennig Society, pp. 3-23. New York Botanical Garden, New York.

Felsenstein, J.
1991   Phylogenetic Inference Package, Version 3.4. University of Washington, Seattle.

Fitch, W. M., and E. Margoliash
1967   Construction of phylogenetic trees. Science 155:279-284.

Fleischer, R. C., S. I. Rothstein, and L. S. Miller
1991   Mitochondrial DNA variation indicates gene flow across a zone of known secondary contact between two subspecies of the Brown-headed Cowbird. Condor 93:185-189.

Fletcher, S. D., and W. S. Moore
1992   Further analysis of allozyme variation in the Northern Flicker, in comparison with mitochondrial DNA variation. Condor 94:988-991.

Frost, J. S., and J. E. Platz
1983   Comparative assessment of modes of reproductive isolation among four species of leopard frogs (*Rana pipiens* complex). Evol. 37:66-78.

Gabriel, K. R.
  1964   A procedure for testing the homogeneity of all sets of means in analysis of variance. Biometrics 20:459-477.

Gabriel, K. R., and R. R. Sokal
  1969   A new statistical approach to geographic variation analysis. Syst. Zool. 18:259-278.

Gaddis, P. K.
  1983   Differential usage of song types by Plain, Bridled and Tufted Titmice. Ornis Scand. 14:16-23.
  1987   Social interactions and habitat overlap between Plain and Bridled Titmice. Southwest. Nat. 32:197-202.

Gambel, W.
  1845   Descriptions of new and little known birds collected in upper California. Proc. Acad. Nat. Sci. Phila. 2:263-266.

Garnett, M. C.
  1981   Body size, its heritability and influence on juvenile survival among Great Tits, *Parus major*. Ibis 123:31-41.

Garrett, K., and J. Dunn
  1981   Birds of Southern California: Status and Distribution. The Artisan Press, Los Angeles.

Gibson, D. D., and B. Kessel
  1989   Geographic variation in the Marbled Godwit and description of an Alaska subspecies. Condor 91:436-443.

Gill, F. B.
  1973   Intra-island variation in the Mascarene White-eye *Zosterops borbonica*. Ornithol. Monogr. 12:1-66.

Gill, F. B., D. H. Funk, and B. Silverin
  1989   Protein relationships among titmice (*Parus*). Wilson Bull. 101:182-197.

Gill, F. B., and B. Slikas
  1992   Patterns of mitochondrial DNA divergence in North American crested titmice. Condor 94:20-28.

Gillespie, J. H.
    1986   Rates of molecular evolution. Ann. Rev. Ecol. Syst. 17:637-665.

Gorman, G. C., and J. Renzi, Jr.
    1979   Genetic distance and heterozygosity estimates in electrophoretic studies: Effects of sample size. Copeia 1979:242-249.

Gosler, A. G., and T. D. Carruthers
    1994   Bill size and niche breadth in the Irish Coal Tit *Parus ater hibernicus*. J. Avian Biol. 25:171-177.

Gould, S. J., and R. F. Johnston
    1972   Geographic variation. Ann. Rev. Ecol. Syst. 3:457-498.

Graybeal, A.
    1993   The phylogenetic utility of cytochrome b: Lessons from bufonid frogs. Mol. Phylog. and Evol. 2:256-269.

Greenwood, P. J.
    1987   Inbreeding, philopatry and optimal outbreeding in birds. In F. Cooke and P. A. Buckley (eds.), Avian Genetics: A Population and Ecological Approach, pp. 207-222. Academic Press, London.

Greenwood, P. J., and P. H. Harvey
    1982   The natal and breeding dispersal of birds. Ann. Rev. Ecol. Syst. 13:1-21.

Griffin, J. R., and W. B. Critchfield
    1972   The distribution of forest trees in California. USDA Forest Service Research Paper PSW-82:1-118.

Grinnell, J.
    1908   The biota of the San Bernardino Mountains. Univ. Calif. Publ. Zool. 5:1-170.
    1923   The present state of our knowledge of the Gray Titmouse in California. Condor 25:135-137.
    1928   A distributional summation of the ornithology of lower California. Univ. Calif. Publ. Zool. 32:1-300.
    1928   Notes on the systematics of West American birds, II. Condor 30:153-156.
    1934   The New Mexico race of Plain Titmouse. Condor 36:251-252.

Grinnell, J., and W. H. Behle
  1937   A new race of titmouse, from the Kern Basin of California. Condor 39:225-226.

Grinnell, J., and A. H. Miller
  1944   The distribution of the birds of California. Pac. Coast Avif. 27:1-617.

Grinnell, J., and T. I. Storer
  1924   Animal Life in the Yosemite. University of California Press, Berkeley.

Grinnell, J., and H. S. Swarth
  1913   An account of the birds and mammals of the San Jacinto area of southern California. Univ. Calif. Publ. Zool. 10:197-406.
  1926   New subspecies of birds (*Penthestes, Beolophus, Psaltriparus, Chamaea*) from the Pacific coast of North America. Univ. Calif. Publ. Zool. 30:163-175.

Groth, J. G.
  1988   Resolution of cryptic species in Appalachian Red Crossbills. Condor 90:745-760.
  1993   Evolutionary differentiation in morphology, vocalizations, and allozymes among nomadic sibling species in the North American Red Crossbill (*Loxia curvirostra*) complex. Univ. Calif. Publ. Zool. 127:1-143.

Grudzien, T. A., W. S. Moore, J. R. Cook, and D. Tagle
  1987   Genic population structure and gene flow in the Northern Flicker (*Colaptes auratus*) hybrid zone. Auk 104:654-664.

Gutiérrez, R. J., R. M. Zink, and S. Y. Yang
  1983   Genic variation, systematic, and biogeographic relationships of some galliform birds. Auk 100:33-47.

Gyllenstein, U. B., and H. A. Erlich
  1988   Generation of single-stranded DNA by the polymerase chain reaction and its application to direct sequencing of the *HLA-DQA* locus. Proc. Natl. Acad. Sci. USA 85:7652-7656.

Hardy, A. C.
  1936   Handbook of Colorimetry. Massachusetts Institute of Technology Press, Cambridge.

Hare, M. P., and G. F. Shields
    1992   Mitochondrial-DNA variation in the polytypic Alaskan Song Sparrow. Auk 109:126-132.

Harris, H., and D. A. Hopkinson
    1976   Handbook of enzyme electrophoresis in human genetics. North Holland Publishing Co., Amsterdam.

Harrison, R. G., and D. M. Rand
    1989   Mosaic hybrid zones and the nature of species boundaries. In D. Otte and J. A. Endler (eds.), Speciation and Its Consequences, pp. 111-133. Sinauer Associates, Sunderland, Mass.

Hedrick, P. W.
    1986   Genetic polymorphism in heterogeneous environments: A decade later. Ann. Rev. Ecol. Syst. 17:535-566.

Hedrick, P. W., M. E. Ginevan, and E. P. Ewing
    1976   Genetic polymorphism in heterogeneous environments. Ann. Rev. Ecol. Syst. 7:1-32.

Hewitt, G. M.
    1988   Hybrid zones—natural laboratories for evolutionary studies. Trends Ecol. Evol. 3:158-167.
    1989   The subdivision of species by hybrid zones. In D. Otte and J. A. Endler (eds.), Speciation and Its Consequences, pp. 85-110. Sinauer Associates, Sunderland, Mass.

Hill, G. E.
    1992   Proximate basis of variation in carotenoid pigmentation in male House Finches. Auk 109:1-12.

Hillis, D. M.
    1984   Misuse and modification of Nei's genetic distance. Syst. Zool. 33:238-240.
    1987   Molecular versus morphological approaches to systematics. Ann. Rev. Ecol. Syst. 18:23-42.
    1991   Discriminating between phylogenetic signal and random noise in DNA sequences. In M. M. Myamota and J. Cracraft (eds.), Phylogenetic Analysis of DNA Sequences, pp. 278-294. Oxford University Press, New York.

Hillis, D. M., and J. P. Heulsenbeck
    1992   Signal, noise, and reliability in molecular phylogenetic analyses. J. Hered. 83:189-195.

Hinde, R. A.
    1952   The behavior of the Great Tit (*Parus major*) and some other related species. Behav. (Suppl. II):1-201.

Hitachi Software Engineering America
    1991   Macintosh DNA and Protein Sequence Input and Analysis System, Version 1.0. San Bruno, Calif.

Hubbard, J. P.
    1973   Avian evolution in the aridlands of North America. The Living Bird 12:155-196.

Huelsenbeck, J. P.
    1991   Tree-length distribution skewness: An indicator of phylogenetic information. Syst. Zool. 40:257-270.

Irwin, D. M., T. D. Kocher, and A. C. Wilson
    1991   Evolution of the cytochrome *b* gene of mammals. J. Mol. Evol. 32:128-144.

James, F. C.
    1970   Geographic size variation in birds and its relationship to climate. Ecol. 51:365-390.
    1982   The ecological morphology of birds: A review. Ann. Zool. Fenn. 19:265-275.
    1983   Environmental component of morphological differences in birds. Science 221:184-186.

Johnson, D. H., M. D. Bryant, and A. H. Miller
    1948   Vertebrate animals of the Providence Mountains area of California. Univ. Calif. Publ. Zool. 48:221-376.

Johnson, G. B.
    1973   Enzyme polymorphism and biosystematics: The hypothesis of selective neutrality. Ann. Rev. Ecol. Syst. 4:93-116.

Johnson, L. S.
    1987   Pattern of song type use for territorial defence in the Plain Titmouse *Parus inornatus*. Ornis Scand. 18:24-32.

Johnson, N. K.
    1963   Biosystematics of sibling species of flycatchers in the *Empidonax hammondii-oberholseri-wrightii* complex. Univ. Calif. Publ. Zool. 66:79-238.
    1974   Molt and age determination in western and yellowish flycatchers. Auk 91:111-131.
    1978   Patterns of avian geography and speciation in the Intermountain Region. Great Basin Nat. Mem. 2:137-159.
    1980   Character variation and evolution of sibling species in the *Empidonax difficilis-flavescens* complex (Aves: Tyrannidae). Univ. Calif. Publ. Zool. 112:1-151.
    1982   Retain subspecies—at least for the time being. Auk 99:605-606.

Johnson, N. K., and A. H. Brush
    1972   Analysis of polymorphism in the Sooty-capped Bush Tanager. Syst. Zool. 21:245-262.

Johnson, N. K., and C. Cicero
    1985   The breeding avifauna of San Benito Mountain, California: Evidence for change over one-half century. West. Birds 16:1-23.
    1986   Richness and distribution of montane avifaunas in the White-Inyo region, California. In C. A. Hall, Jr. and D. J. Young (eds.), Natural History of the White-Inyo Range, Eastern California and Western Nevada and High Altitude Physiology, pp. 137-159. University of California White Mountain Research Station Symposium, vol. 1. Bishop, Calif.
    1991   Mitochondrial DNA sequence variability in two species of sparrows of the genus *Amphispiza*. Acta Congr. Intern. Ornithol. 20:600-610.

Johnson, N. K., and J. A. Marten
    1988   Evolutionary genetics of flycatchers. II. Differentiation in the *Empidonax difficilis* complex. Auk 105:177-191.
    1992   Macrogeographic patterns of morphometric and genetic variation in the Sage Sparrow complex. Condor 94:1-19.

Johnson, N. K., R. M. Zink, G. F. Barrowclough, and J. A. Marten
    1984   Suggested techniques for modern avian systematics. Wilson Bull. 96:543-560.

Johnston, R. F.
    1969   Character variation and adaptation in the European sparrows. Syst. Zool. 18:206-231.
    1972   Ecologic differentiation in North American birds. Univ. Ark. Mus. Occas. Paper no. 4:101-132.

Johnston, R. F., and R. K. Selander
    1964   House Sparrows: Rapid evolution of races in North America. Science 144:548-550.
    1971   Evolution in the House Sparrow. II. Adaptive differentiation in North American populations. Evol. 25:1-28.

Judd, D. B.
    1933   The 1931 I.C.I. standard observer and coordinate system for colorimetry. J. Opt. Soc. Amer. 23:359-374.

Kat, P. W.
    1985   Historical evidence for fluctuation in levels of hybridization. Evol. 39:1164-1169.

Kimura, M.
    1979   The neutral theory of molecular evolution. Sci. Am. 241:98-126.
    1980   A simple method of estimating evolutionary rate of base substitutions through comparative studies of nucleotide sequences. J. Mol. Evol. 16:111-120.
    1982   The neutral theory as a basis for understanding the mechanism of evolution and variation at the molecular level. In M. Kimura (ed.), Molecular Evolution, Protein Polymorphism and the Neutral Theory, pp. 3-56. Scientific Societies Press, Tokyo.

King, J. E., and T. R. Van Devender
    1977   Pollen analysis of fossil packrat middens from the Sonoran Desert. Quat. Res. 8:191-204.

King, T. J., Jr.
    1976   Late Pleistocene-early Holocene history of coniferous woodlands in the Lucerne Valley region, Mohave Desert, California. Great Basin Nat. 36:227-238.

Kingery, H. E.
    1981   Mountain West region. Amer. Birds 35:963-966.

1982   Mountain West. Amer. Birds 36:1000-1003.

Kocher, T. D., W. K. Thomas, A. Meyer, S. V. Edwards, S. Pääbo, F. X. Villablanca, and A. C. Wilson
1989   Dynamics of mitochondrial DNA evolution in animals: Amplification and sequencing with conserved primers. Proc. Natl. Acad. Sci. USA 86:6196-6200.

Kocher, T. D., and T. J. White
1989   Evolutionary analysis via PCR. In H. A. Erlich (ed.), PCR Technology: Principles and Applications for DNA Amplification, pp. 137-147. Stockton Press, New York.

Kumar, S., K. Tamura, and M. Nei
1993   Molecular Evolutionary Genetics Analysis, Version 1.01. Institute of Molecular Evolutionary Genetics, Pennsylvania State University, University Park.

Lack, D.
1954   The Natural Regulation of Animal Numbers. Clarendon Press, Oxford.

Lara, M., M. A. Bogan, and R. Cerqueira
1992   Sex and age components of variation in *Proechimys cuvieri* (Rodentia: Echimyidae) from northern Brazil. Proc. Biol. Soc. Wash. 105:882-893.

Larson, A., D. B. Wake, and K. P. Yanev
1984   Measuring gene flow among populations having high levels of genetic fragmentation. Genet. 106:293-308.

Leberg, P. L.
1992   Effects of population bottlenecks on genetic diversity as measured by allozyme electrophoresis. Evol. 46:477-494.

Leskinen, P. H.
1975   Occurrence of oaks in late Pleistocene vegetation in the Mojave Desert of Nevada. Madroño 23:234-235.

Lessa, E. P.
1990   Multidimensional analysis of geographic genetic structure. Syst. Zool. 39:242-252.

Levene, H.
1949   On a matching problem arising in genetics. Ann. Math. Stat. 20:91-94.

Linsdale, J. M.
1938a   Geographic variation in some birds in Nevada. Condor 40:36-38.
1938b   Environmental responses of vertebrates in the Great Basin. Amer. Midl. Nat. 19:1-216.
1938c   The birds of Nevada. Pac. Coast Avif. 23:5-145.

Littlejohn, M. J., and G. F. Watson
1985   Hybrid zones and homogamy in Australian frogs. Ann. Rev. Ecol. Syst. 16:85-112.

Lougheed, S. C., T. W. Arnold, and R. C. Bailey
1991   Measurement error of external and skeletal variables in birds and its effect on principal components. Auk 108:432-436.

Mack, A. L., F. B. Gill, R. Colburn, and C. Spolsky
1986   Mitochondrial DNA: A source of genetic markers for studies of similar passerine bird species. Auk 103:676-681.

Maniatis, T., E. F. Fritsch, and J. Sambrook
1982   Molecular Cloning: A Laboratory Manual. Cold Spring Harbor Laboratory, Cold Spring Harbor, New York.

Manly, B. F. J.
1986   Multivariate Statistical Methods: A Primer. Chapman and Hall, London.

Mantel, N.
1967   The detection of disease clustering and a general regression approach. Cancer Res. 27:209-220.

Marshall, J. T., Jr.
1957   Birds of pine-oak woodland in southern Arizona and adjacent Mexico. Pac. Coast Avif. 32:5-125.

Marten, J. A., and N. K. Johnson
1986   Genetic relationships of North American cardueline finches. Condor 88:409-420.

Martin, A. C., H. S. Zim, and A. L. Nelson
	1951	American Wildlife and Plants. Dover, New York.

Martin, J.
	1991	Patterns and significance of geographical variation in the Blue Tit (*Parus caeruleus*). Auk 108:820-832.

Martin, J., and J. Pitochelli
	1991	Relation of within-population phenotypic variation with sex, season, and geography in the Blue Tit. Auk 108:833-841.

Matthysen, E.
	1990	Nonbreeding social organization in *Parus*. Current Ornithol. 7:209-249.

Mayr, E.
	1942	Systematics and the Origin of Species. Columbia University Press, New York.
	1963	Animal Species and Evolution. Columbia University Press, New York.

Mayr, E., and R. J. O'Hara
	1986	The biogeographic evidence supporting the Pleistocene forest refuge hypothesis. Evol. 40:55-67.

McCarten, N., and T. R. Van Devender
	1988	Late Wisconsin vegetation of Robber's Roost in the western Mojave Desert, California. Madroño 35:226-237.

McCauley, D. E.
	1993	Genetic consequences of extinction and recolonization in fragmented habitats. In P. M. Kareiva, J. G. Kingsolver and R. B. Huey (eds.), Biotic Interactions and Global Change, pp. 217-233. Sinauer Associates, Sunderland, Mass.

McDonald, J. F.
	1983	The molecular basis of adaptation: A critical review of relevant ideas and observations. Ann. Rev. Ecol. Syst. 14:77-102.

McDonald, M. A., and M. H. Smith
	1990	Speciation, heterochrony, and genetic variation in Hispanolian Palm-Tanagers. Auk 107:707-717.

McKitrick, M. C., and R. M. Zink
    1988   Species concepts in ornithology. Condor 90:1-14.

Mengel, R. M.
    1964   The probable history of species formation in some northern wood warblers (Parulidae). The Living Bird 3:9-43.

Michener, H., and J. R. Michener
    1940   The molt of House Finches of the Pasadena region, California. Condor 42:140-153.

Miller, A. H.
    1941   A review of centers of differentiation for birds in the western Great Basin. Condor 43:257-267.
    1946   Endemic birds of the Little San Bernardino Mountains, California. Condor 48:75-79.

Miller, A. H., and R. C. Stebbins
    1964   The Lives of Desert Animals in Joshua Tree National Monument. University of California Press, Berkeley.

Miller, R. F., and P. E. Wigand
    1994   Holocene changes in semiarid pinyon-juniper woodlands. Bioscience 44:465-474.

Mitton, J. B.
    1994   Molecular approaches to population biology. Ann. Rev. Ecol. Syst. 25:45-69.

Moen, S. M.
    1991   Morphological and genetic variation among breeding colonies of the Atlantic Puffin (*Fratercula arctica*). Auk 108:755-763.

Moore, W. S., and D. B. Buchanan
    1985   Stability of the Northern Flicker hybrid zone in historical times: Implications for adaptive speciation theory. Evol. 39:135-151.

Moore, W. S., and R. A. Dolbeer
    1989   The use of banding recovery data to estimate dispersal rates and gene flow in avian species: Case studies in the Red-winged Blackbird and Common Grackle. Condor 91:242-253.

Moore, W. S., J. H. Graham, and J. Price
  1991   Mitochondrial DNA variation in the Northern Flicker (*Colaptes auratus*, Aves). Mol. Biol. Evol. 8:327-344.

Moore, W. S., and W. D. Koenig
  1986   Comparative reproductive success of Yellow-shafted, Red-shafted, and hybrid flickers across a hybrid zone. Auk 103:42-51.

Moritz, C., T. E. Dowling, and W. M. Brown
  1987   Evolution of animal mitochondrial DNA: Relevance for population biology and systematics. Ann. Rev. Ecol. Syst. 18:269-292.

Mousseau, T. A., and D. A. Roff
  1989   Adaptation to seasonality in a cricket: Patterns of phenotypic and genotypic variation in body size and diapause expression along a cline in season length. Evol. 43:1483-1496.

Murphy, E. C.
  1985   Bergmann's rule, seasonality, and geographic variation in body size of House Sparrows. Evol. 39:1327-1334.

Nei, M.
  1975   Molecular Population Genetics and Evolution. North Holland Publishing Co., Amsterdam.
  1978   Estimation of average heterozygosity and genetic distance from a small number of individuals. Genet. 89:583-590.

Nei, M., T. Maruyama, and R. Chakraborty
  1975   The bottleneck effect and genetic variability in populations. Evol. 29:1-10.

Nei, M., F. Tajima, and Y. Tateno
  1983   Accuracy of estimated phylogenetic trees from molecular data. II. Gene frequency data. J. Mol. Evol. 19:153-170.

Nevo, E., and A. Beiles
  1989   Genetic diversity in the desert: Patterns and testable hypotheses. J. Arid Environ. 17:241-244.

Nores, M.
  1992   Bird speciation in subtropical South America in relation to forest expansion and retraction. Auk 109:346-357.

Norris, R. A.
    1958   Comparative biosystematics and life history of the nuthatches *Sitta pygmaea* and *Sitta pusilla*. Univ. Calif. Publ. Zool. 56:119-300.

Oberholser, H. C.
    1932   Descriptions of new birds from Oregon, chiefly from the Warner Valley region. Cleveland Mus. Nat. Hist. Sci. Publ. 4:1-12.

Ohta, T.
    1992   The nearly neutral theory of molecular evolution. Ann. Rev. Ecol. Syst. 23:263-286.

O'Neill, J. P.
    1982   The subspecies concept in the 1980's. Auk 99:609-612.

Otte, D., and J. A. Endler (eds.)
    1989   Speciation and its Consequences. Sinauer Associates, Sunderland, Mass.

Owenby, J. R., and D. S. Ezell
    1992   Monthly station normals of temperature, precipitation, and heating and cooling degree days, 1961-1990. Climatography of the United States no. 81 (by state). National Oceanic and Atmospheric Administration, National Climatic Data Center, Asheville, N.C.

Parkes, K. C.
    1982   Subspecific taxonomy: Unfashionable does not mean irrelevant. Auk 99:596-598.

Patten, M. A., P. Unitt, R. A. Erickson, and K. F. Campbell
    1995   Fifty years since Grinnell and Miller: Where is California ornithology headed? West. Birds 26:54-64.

Patton, J. L., and P. V. Brylski
    1987   Pocket gophers in alfalfa fields: Causes and consequences of habitat-related body size variation. Amer. Nat. 130:493-506.

Patton, J. L., and M. F. Smith
    1990   The evolutionary dynamics of the pocket gopher *Thomomys bottae*, with emphasis on California populations. Univ. Calif. Publ. Zool. 123:1-161.
    1992   mtDNA phylogeny of Andean mice: A test of diversification across ecological gradients. Evol. 46:174-183.

Pavlik, B. M., P. C. Muick, S. Johnson, and M. Popper
    1991   Oaks of California. Cachuma Press, Los Olivos, Calif.

Payne, R. B.
    1986   Bird songs and avian systematics. Current Ornithol. 3:87-126.

Paynter, R. A., Jr. (ed.)
    1967   Check-list of Birds of the World: A Continuation of the Work of James L. Peters, vol. 12. The Heffernan Press, Worcester, Mass.

Peterson, A. T.
    1990   Birds of Eagle Mountain, Joshua Tree National Monument, California. West. Birds 21:127-135.
    1992   Phylogeny and rates of molecular evolution in the *Aphelocoma* jays (Corvidae). Auk 109:133-147.

Phillips, A. R., J. Marshall, and G. Monson
    1964   The Birds of Arizona. University of Arizona Press, Tucson.

Pielou, E.C.
    1991   After the Ice Age: The Return of Life to Glaciated North America. University of Chicago Press, Chicago.

Pitelka, F. A.
    1945   Pterylography, molt, and age determination of American jays of the genus *Aphelocoma*. Condor 47:229-260.

Pitochelli, J.
    1990   Plumage, morphometric, and song variation in Mourning (*Oporornis philadelphia*) and MacGillivray's (*O. tolmiei*) Warblers. Auk 107:161-171.

Porter, S. C. (ed.)
    1983   Late-Quaternary Environments of the United States, vol. 1. The Late Pleistocene. University of Minnesota Press, Minneapolis.

Power, D. M.
    1970   Geographic variation of Red-winged Blackbirds in central North America. Univ. Kansas Publ. Mus. Nat. Hist. 19:1-83.

Prager, E. M., and A. C. Wilson
    1978   Construction of phylogenetic trees for proteins and nucleic acids: Comparison of alternative matrix methods. J. Mol. Evol. 11:129-142.

Price, J. B.
    1936   The family relations of the Plain Titmouse. Condor 38:23-28.

Pulliam, H. R.
    1988   Sources, sinks, and population regulation. Amer. Nat. 132:652-661.

Pyle, P., S. N. G. Howell, R. P. Yunick, and D. F. DeSante
    1987   Identification Guide to North American Passerines. Slate Creek Press, Bolinas, Calif.

Quinn, T. W.
    1992   The genetic legacy of Mother Goose: Phylogeographic patterns of Lesser Snow Goose *Chen caerulescens caerulescens* maternal lineages. Mol. Ecol. 1:105-117.

Rand, A. L.
    1948   Glaciation, an isolating factor in speciation. Evol. 2:314-321.

Richmond, C. W.
    1902   *Parus inornatus griseus* renamed. Proc. Biol. Soc. Wash. 15:155.

Riddle, B. R., and R. L. Honeycutt
    1990   Historical biogeography in North American arid regions: An approach using mitochondrial DNA phylogeny in grasshopper mice (genus *Onychomys*). Evol. 44:1-15.

Ridgway, R.
    1882   Descriptions of some new North American birds. Proc. U.S. Natl. Mus.5:343-346.
    1883   Descriptions of some new birds from Lower California, collected by Mr. L. Belding. Proc. U.S. Nat. Mus. 6:154-156.
    1903   Descriptions of new genera, species and subspecies of American birds. Proc. Biol. Soc. Wash. 16:105-112.
    1904   The birds of North and Middle America. U.S. Natl. Mus. Bull. 50:387-392.

Robbins, M. B., M. J. Braun, and E. A. Tobey
  1986   Morphological and vocal variation across a contact zone between the chickadees *Parus atricapillus* and *P. carolinensis*. Auk 103:655-666.

Rockwell, R. F., and G. F. Barrowclough
  1987   Gene flow and the genetic structure of populations. In F. Cooke and P. A. Buckley (eds.), Avian Genetics: A Population and Ecological Approach, pp. 223-255. Academic Press, London.

Rogers, J. S.
  1972   Measures of genetic similarity and genetic distance. Univ. Texas Stud. Genet. 7:145-153.

Rogers, T. H.
  1985   Northern Rocky Mountain-Intermountain region. Amer. Birds 39:938-941.

Rohlf, F. J.
  1988   NTSYS-pc Numerical Taxonomy and Multivariate Analysis System, Version 1.50. Applied Biostatistics, Setauket, New York.

Ruffner, J. A.
  1985   Climates of the States, 3rd ed. Gale Research Co., Detroit.

Sage, R. D., and J. O. Wolff
  1986   Pleistocene glaciations, fluctuating ranges, and low genetic variability in a large mammal (*Ovis dalli*). Evol. 40:1092-1095.

Saitou, N., and M. Nei
  1987   The neighbor-joining method: A new method for reconstructing phylogenetic trees. Mol. Biol. Evol. 4:406-425.

Salt, G. W.
  1952   The relation of metabolism to climate and distribution in three finches of the genus *Carpodacus*. Ecol. Monogr. 22:121-152.

Sanger, P., S. A. Nicklen, and A. R. Coulsen
  1977   DNA sequencing with chain-terminating inhibitors. Proc. Natl. Acad. Sci. USA 74:5463-5467.

Schnell, G. D., D. J. Watt, and M. E. Douglas
  1985   Statistical comparison of proximity matrices: Applications in animal behavior. Anim. Behav. 33:239-253.

Selander, R. K.
  1971   Systematics and speciation in birds. In D. S. Farner and J. R. King (eds.), Avian Biology, vol. 1, pp. 57-147. Academic Press, New York.

Selander, R. K., M. H. Smith, S. Y. Yang, W. E. Johnson, and J. B. Gentry
  1971   Biochemical polymorphism and systematics in the genus *Peromyscus*. I. Variation in the old-field mouse (*Peromyscus polionotus*). Univ. Texas Stud. Genet. 6:49-90.

Shaffer, H. B., J. M. Clark, and F. Kraus
  1991   When molecules and morphology clash: A phylogenetic analysis of the North American ambystomatid salamanders (Caudata: Ambystomatidae). Syst. Zool. 40:284-303.

Sheldon, F. H., B. Slikas, M. Kinnarney, F. B. Gill, E. Zhao, and B. Silverin
  1992   DNA-DNA hybridization evidence of phylogenetic relationships among major lineages of *Parus*. Auk 109:173-185.

Shields, G. F., and K. M. Helm-Bychowski
  1988   Mitochondrial DNA of birds. Current Ornithol. 5:273-295.

Shields, G. F., and A. C. Wilson
  1987a   Subspecies of the Canada Goose (*Branta canadensis*) have distinct mitochondrial DNA's. Evol. 41:662-666.
  1987b   Calibration of mitochondrial DNA evolution in geese. J. Mol. Evol. 24:212-217.

Simpson, G. G.
  1944   Tempo and Mode of Evolution. Columbia University Press, New York.

Slatkin, M.
  1980   The distribution of mutant alleles in a subdivided population. Genet. 95:503-523.
  1981   Estimating levels of gene flow in natural populations. Genet. 99:323-335.
  1985a   Gene flow in natural populations. Ann. Rev. Ecol. Syst. 16:393-430.
  1985b   Rare alleles as indicators of gene flow. Evol. 39:53-65.

1987   Gene flow and the geographic structure of natural populations. Science 236:787-792.

1989   Detecting small amounts of gene flow from phylogenies of alleles. Genetics 121:609-612.

Smith, J. N. M., and A. A. Dhondt
1980   Experimental confirmation of heritable morphological variation in a natural population of Song Sparrows. Evol. 34:1155-1158.

Smith, M. F., and J. L. Patton
1991   Variation in mitochondrial cytochrome b sequence in natural populations of South American akodontine rodents (Muridae: Sigmodontinae). Mol. Biol. Evol. 8:85-103.

Smith, M. F., W. K. Thomas, and J. L. Patton
1992   Mitochondrial DNA-like sequence in the nuclear genome of an akodontine rodent. Mol. Biol. Evol. 9:204-215.

Sneath, P. H. A., and R. R. Sokal
1973   Numerical Taxonomy. W. H. Freeman, San Francisco.

Sokal, R. R.
1979   Testing statistical significance of geographic variation patterns. Syst. Zool. 28:227-232.

Sokal, R. R., and F. J. Rohlf
1969   Biometry, 1st ed. W. H. Freeman, San Francisco.

SPSS Inc.
1986   SPSS[X] User's Guide, 2nd ed. SPSS Inc., Chicago.

St. Andre, G., H. A. Mooney, and R. D. Wright
1965   The pinyon woodland zone in the White Mountains of California. Amer. Midl. Nat. 73:225-239.

Storer, R. W.
1982   Subspecies and the study of geographic variation. Auk 99:599-601.

Storer, R. W., and D. A. Zimmerman
 1959   Variation in the Blue Grosbeak (*Guiraca caerulea*) with special reference to the Mexican populations. Univ. Mich. Mus. Zool. Occas. Papers 609:1-13.

Straney, D. O., and J. L. Patton
 1980   Phylogenetic and environmental determinants of geographic variation of the pocket mouse *Perognathus goldmani* Osgood. Evol. 34:888-903.

Swofford, D. L.
 1981   On the utility of the distance Wagner procedure. In V. A. Funk and D. R. Brooks (eds.), Advances in Cladistics: Proceedings of the First Meeting of the Willi Hennig Society, pp. 25-43. New York Botanical Garden, New York.
 1993   Phylogenetic Analysis Using Parsimony, Version 3.1.1. Laboratory of Molecular Systematics, Smithsonian Institution, Washington.

Swofford, D. L., and G. J. Olson
 1990   Phylogeny reconstruction. In D. M. Hillis and C. Moritz (eds.), Molecular Systematics, pp. 411-501. Sinauer Associates, Sunderland, Mass.

Swofford, D. L., and R. K. Selander
 1981   BIOSYS-1: A FORTRAN program for the comprehensive analysis of genetic data in population genetics and systematics. J. Hered. 72:281-283.

Tamura, K., and M. Nei
 1993   Estimation of the number of nucleotide substitutions in the control region of mitochondrial DNA in humans and chimpanzees. Mol. Biol. and Evol. 10:512-526.

Tegelström, H., H. P. Gelter, and M. Jaarola
 1990   Variation in Pied Flycatcher (*Ficedula hypoleuca*) mitochondrial DNA. Auk 107:730-736.

Temple, S. A.
 1972   Systematics and evolution of the North American merlins. Auk 89:325-338.

Templeton, A. R.
 1980   Modes of speciation and inferences based on genetic distances. Evol. 34:719-729.

1981   Mechanisms of speciation: A population genetic approach. Ann. Rev. Ecol. Syst. 12:23-48.

Thomas, W. K., S. Pääbo, F. X. Villablanca, and A. C. Wilson
1990   Spatial and temporal continuity of kangaroo rat populations shown by sequencing mitochondrial DNA from museum specimens. J. Mol. Evol. 31:101-112.

Thompson, R. S., and J. I. Mead
1982   Late Quaternary environments and biogeography in the Great Basin. Quat. Res. 17:39-55.

Thorpe, J. P.
1982   The molecular clock hypothesis: Biochemical evolution, genetic differentiation and systematics. Ann. Rev. Ecol. Syst. 13:139-168.

Thorpe, R. S.
1976   Biometric analysis of geographic variation and racial affinities. Biol. Rev. 51:407-452.
1983   A review of the numerical methods for recognising and analysing racial differentiation. In J. Felsenstein (ed.), Numerical Taxonomy, pp. 404-423. Springer-Verlag, Berlin.

Van Devender, T. R.
1977   Holocene woodlands in the southwestern deserts. Science 198:189-192.

Van Devender, T. R., and W. G. Spaulding
1979   Development of vegetation and climate in the southwestern United States. Science 204:701-710.

Van Noordwijk, A.
1987   Quantitative ecological genetics of Great Tits. In F. Cooke and R. A. Buckley (eds.), Avian Genetics: A Population and Ecological Approach, pp. 363-380. Academic Press, London.

Van Wagner, C. E., and A. J. Baker
1986   Genetic differentiation in populations of Canada Geese (*Branta canadensis*). Can. J. Zool. 64:940-947.

Vawter, L., and W. M. Brown
   1986   Nuclear and mitochondrial DNA comparisons reveal extreme rate variation in the molecular clock. Science 234:194-196.

Via, S., and R. Lande
   1985   Genotype-environment interaction and the evolution of phenotypic plasticity. Evol. 39:505-522.

Wade, M. J., and D. E. McCauley
   1988   Extinction and recolonization: Their effects on the genetic differentiation of local populations. Evol. 42:995-1005.

Walsh, J. B.
   1982   Rate of accumulation of reproductive isolation by chromosome rearrangements. Amer. Nat. 120:510-532.

Wells, P. V.
   1966   Late Pleistocene vegetation and degree of pluvial climatic change in the Chihuahuan Desert. Science 153:970-975.
   1979   An equable glaciopluvial in the West: Pleniglacial evidence of increased precipitation on a gradient from the Great Basin to the Sonoran and Chihuahuan deserts. Quat. Res. 12:311-325.
   1983   Paleobiogeography of montane islands in the Great Basin since the last glaciopluvial. Ecol. Monogr. 53:341-382.

Wells, P. V., and R. Berger
   1967   Late Pleistocene history of coniferous woodland in the Mohave Desert. Science 155:1640-1647.

Wells, P. V., and C. D. Jorgensen
   1964   Pleistocene wood rat middens and climatic change in Mohave Desert: A record of juniper woodlands. Science 143:1171-1174.

West-Eberhard, M. J.
   1989   Phenotypic plasticity and the origins of diversity. Ann. Rev. Ecol. Syst. 20:249-278.

White, T. J., N. Arnheim, and H. A. Erlich
   1989   The polymerase chain reaction. Trends in Genet. 5:185-189.

Whitlock, M. C., and D. E. McCauley
    1990   Some population genetic consequences of colony formation and extinction: Genetic correlations within founding groups. Evol. 44:1717-1724.

Wiedenfeld, D. A.
    1991   Geographical morphology of male Yellow Warblers. Condor 93:712-723.

Wiens, J. A. (ed.)
    1982.   Forum: Avian subspecies in the 1980's. Auk 99:593-615.

Wilbur, S. R.
    1987   Birds of Baja California. University of California Press, Berkeley.

Wilson, A. C., R. L. Cann, S. M. Carr, M. George, U. B. Gyllenstein, K. M. Helm-Bychowski, R. G. Higuchi, S. R. Palumbi, E. M. Prager, R. D. Sage, and M. Stoneking
    1985   Mitochondrial DNA and two perspectives on evolutionary genetics. Biol. J. Linn. Soc. 26:375-400.

Wilson, E. O., and W. L. Brown, Jr.
    1953   The subspecies concept and its taxonomic application. Syst. Zool. 2:97-111.

Wright, H. E., Jr. (ed.)
    1983   Late-Quaternary Environments of the United States, vol. 2. The Holocene. University of Minnesota Press, Minneapolis.

Wright, S.
    1951   The genetical structure of populations. Ann. Eugen. 15:323-354.
    1965   The interpretation of population structure by F-statistics with special regard to systems of mating. Evol. 19:395-420.
    1978   Evolution and the Genetics of Natural Populations, vol. 4. Variability Within and Among Natural Populations. University of Chicago Press, Chicago.

Zink, R. M.
    1986   Patterns and evolutionary significance of geographic variation in the Schistacea group of the Fox Sparrow (*Passerella iliaca*). Ornithol. Monogr. 40:1-119.
    1989   The study of geographic variation. Auk 106:157-160.
    1991   The geography of mitochondrial DNA variation in two sympatric sparrows. Evol. 45:329-339.

Zink, R. M., D. F. Lott, and D. W. Anderson

    1987   Genetic variation, population structure, and evolution of California Quail. Condor 89:395-405.

Zink, R. M., and J. V. Remsen,

    1986   Evolutionary processes and patterns of geographic variation in birds. Current Ornithol. 4:1-69.

Zink, R. M., W. L. Rootes, and D. L. Dittmann

    1991   Mitochondrial DNA variation, population structure, and evolution of the Common Grackle (*Quiscalus quiscula*). Condor 93:318-329.